The Flavones

THE AUTHOR

Dr. Shankar Nagnath Ipper is working as an Assistant Professor in Department of Chemistry at Sunderrao Solanke Mahavidyalaya, Majalgaon Dist. Beed. He has completed his M.Sc. (Organic Chemistry) from Yeshwant College Nanded affiliated to Swami Ramanand Teerth Marathwada University Nanded in 2004. He has qualified SET in August 2006. He has completed his Ph.D. under the guidance of Dr. G. K. Kakade, Research Guide, ACS College Kille-Dharur, Dist Beed affiliated to Dr. B. A. M. University, Aurangabad. He has more than 13 years experience of teaching. He has published more than 20 research papers in various reputed International Journals and also attended more than 50 Conferences, Seminar and Workshop. Dr. Shankar Ipper has honoured with "Best Researcher Award" International Best Scientist Awards, VDGOOD Professional Association, New Delhi on 18 & 19th Jan-2020.

The Flavones

Dr. Shankar Nagnath Ipper
Assistant Professor
Head, Department of Chemistry,
Sunderrao Solanke Mahavidyalaya,
Majalgaon Dist. Beed (M.S.)

2021

Scholars World

A Division of

Astral International Pvt. Ltd.

New Delhi – 110 002

Cataloging in Publication Data--DK
Courtesy: D.K. Agencies (P) Ltd. <docinfo@dkagencies.com>

Ipper, Shankar Nagnath, author.
The flavones / Dr. Shankar Nagnath Ipper.
pages cm
Includes bibliographical references.
ISBN 9789354616020 (HB)

1. Flavonoids. I. Title.

LCC QK898.F5I67 2021 | DDC 547.869 23

Published by : **Scholars World**
A Division of
Astral International Pvt. Ltd.
– ISO 9001:2015 Certified Company –
4736/23, Ansari Road, Darya Ganj,
New Delhi-110 002
Ph. 011-43549197, 23278134
E-mail: info@astralint.com
Website: www.astralint.com

Respectfully
Dedicated
To
My Beloved Parents
Shri. Nagnathrao Ipper
Sou. Sumanbai Ipper

Acknowledgment

It gives me immense pleasure to express my deep sense of gratitude and indebtedness to my research guide. Respected **Dr. G. K. Kakade**, Department of chemistry, Arts, Commerce & Science College Kille-Dharur for his inspiring guidance stimulating discussions and constant encouragement throughout research and writing work.

I am very much thankful to **MSP Mandals President Shri Prakash Solanke, Secretary Shri Satish Chavan** and Principal **Dr.V.P. Pawar**, Sunderrao Solanke Mahavidyalaya, Majalgaon, for inspiration me to publish my research work.

I also express my sincere thanks to the **Dr. A.A. Kachare, Dr. A. M. Zine, Dr. N. K. Raut, Dr. G. K. Sanap,** to encourage me to publish research work.

My heartfelt thank are due to my expected father in law and mother in law **Shri. Keshav V. Kagne** and **Sou. Chhaya K. Kagne** for their deviation during my book writing.

I am very much thankful to my Wife **Sou.** ***Ashwini S. Ipper*** for his persistence support and encouragement during the progress of the book writing.

I am specially thanks to my sweet daughter **Bhakti S. Ipper** for constant encouragement throughout my book writing.

Dr. Shankar Nagnath Ipper

Preface

The present book attempts to give to the students, learning natural products of organic compounds the benefit of thirteen years experience in teaching and research of this subject by author. This book is very useful to new researcher how to start research and gives idea how to done research in stepwise way. The subject matter presented in this book is knowledge gained through it will permit identification of an unknown flavones and their metal complexes.

The present work was carried out to understand the structural and biological properties of flavones and their metal complexes. Flavone ligands are synthesized by using chalcones of 2-furaldehyde and 5-methyl-2-furaldehyde by using DMSO/I2 as an oxidizing agent and have been used for the first time in the synthesis and studies on structural aspects of transition metal complexes. The overall chapters are summarized below.

CHAPTER-1 INTRODUCTION

In this chapter the basic concepts, concerned flavones, their metal complexes and applicational importance has been discussed. A brief review of literature on previous studies on metal complexes of and flavones has been surveyed in brief. At the end of this chapter the aim and scope of present investigation has been outlined.

CHAPTER-2 SYNTHESIS AND CHARACTERIZATION OF FLAVONES AND THEIR METAL COMPLEXES

This chapter deals with the synthesis of flavones and their characterization by FTIR spectroscopy. General method for the synthesis of Mn(II), Co(II), Ni(II), Cu(II), Zn(II) and Cd(II) complexes have been described.

The procedure and techniques employed for characterization of Flavones by elemental analysis, UV- visible spectra, mass spectra, IR, 1H-NMR, spectroscopy have been included. The physical parameters of the flavones are discussed. Analytical data and mass spectra of flavones and metal complexes have been discussed.

CHAPTER-3 EXPERIMENTAL TECHNIQUES

This chapter describes various experimental techniques and their role in the characterization of flavones and metal complexes. The experimental details of the measurements of magnetic susceptibility, solution conductivity, scanning of 1H-NMR, electronic absorption spectra, infrared spectra, mass spectra, X-ray diffraction spectra and TG-DTA of metal complexes are given.

CHAPTER- 4 INFRARED SPECTROSCOPY

This chapter deals with the results and discussion pertaining to infrared absorption spectra of flavones and their metal complexes. The infrared spectral study of flavones and their metal complexes with respect to assignment of band frequencies for different groups involved in complex formation.

CHAPTER- 5 ELEMETAL ANALYSIS, SOLUTION CONDUCTIVITY, MAGNETIC SUSCEPTIBILITY AND ELECTRONIC ABSORPTION SPECTRAL STUDIES OF METAL COMPLEXES

In this chapter results and discussion of experimental data of magnetic susceptibility, electronic absorption spectra and solution conductivity of representative metal complexes of flavones are discussed. The experimental data on conductivity, magnetic susceptibility, electronic spectra, and the electronic absorption spectra of the metal complexes are discussed.

CHAPTER- 6 THERMAL STUDIES OF METAL COMPLEXES

The chapter includes discussion on thermal behavior of metal complexes on the basis of simultaneous TG-DTA curves and thermodynamic parameters.

CHAPTER- 7 X-RAY DIFFRACTION SPECTRAL STUDIES OF METAL COMPLEXES

X-ray diffraction measurements of the representative metal complexes are discussed. The unit cell data and crystal lattice of metal complexes calculated from strong reflections in the XRD pattern by using computer program are presented and discussed with respect to Z values, density, porosity, unit cell volume, standard deviation, crystal system and probable space group.

CHAPTER- 8 BIOLOGICAL STUDIES OF, FLAVONES AND THEIR METAL COMPLEXES

This chapter deals with antibacterial and antifungal activity of flavones and their metal complexes. It is divided into two sections. Section first includes the aspects of bacteria, experimental procedure for the measurement of antibacterial activity of flavones and their metal complexes invitro against Gram positive pathogens Staphylococcus aureus (ATCC6538), Bacillus subtilis(ATCC6633), Bacillus cereus(ATCC14579), Bacillus megaterium (ATCC2326), Micrococcus glutamicum and Gram negative pathogens Escherishia coli (ATCC8739), Salmonella typhi (ATCC9207), Shigella boydii (ATCC12034), Enterobacter aerogenes (ATCC13048), Pseudomonas aerogenosa (ATCC9027), Salmonella abony (NCTC6017) by agar cup method at 1% concentration. The inhibitory zone and data are presented.

Section second describes introduction to fungi, antifungal activity of flavones and their metal complexes invitro against Fungal pathogens Aspergillus niger(ATCC16404), Saccharomyces cereviseae (ATCC9763), Candida albicans (ATCC10231) species by poison plate method.

CHAPTER-9 SUMMARY AND CONCLUSIONS

This chapter includes the Summary and Conclusions. Lastly the book provides a large no of references at the end of chapters.

Dr. Shankar Ipper

Contents

CHAPTER 1

Introduction

1.1. Introduction of Flavones:

Flavonoids are also known as plant pigments or co-pigments. Flavonoids are the largest group of naturally occurring phenolic compounds which occur in different plant parts both in a free state and as glycosides. Flavonoids are found to occur in different parts of the plant roots, barks, wood, leaves, flowers, fruits, and seeds. Flavonoids are synthesized in all parts of the plants. They play role in providing color, fragrance and taste to the fruits, flowers and seeds, which makes them attractants for insects, birds or mammals, which aid in pollen or seed transmission.[1] These compounds are grouped together as chalcones, di-hydro-chalcones, aurones, flavones, flavonones, isoflavonones, 2,3,di-hydroflavonols, 3,4,diols, anthocyanidins and catechins etc.

The term flavonoids are generally used to describe a broad collection of natural products that includes a C_6-C_3-C_6 Carbon frame work or more specifically phenyl benzopyran functionality.

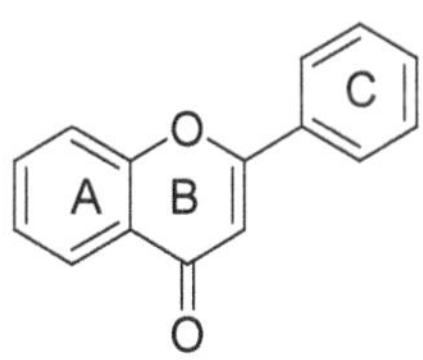

Fig.1.1. The Structure of Flavone.

The name flavone for the unknown 2-phenyl-benzopyranes was first suggested by Von Kostanecki and Tombar. The systematic investigation of natural flavones begins with the studies initiated by Kostanecki, Herzig and Perkin.

The flavones are generally known as anthoxanhins and are widely distributed yellow plant pigments. These anthoxanhins also occur as colorless glycosides in the white corollas of several flowers which on treatment with ammonia vapours turns yellow. Flavonoids are of increasing interest due to their health benefits they posses antioxidant activity, the ability to scavenge active oxygen species and acts as metal chelators.[2-4]

Flavonoids represent one of the largest groups of natural products. In addition to the various functions of flavonoids in plants, their widespread distribution in nature, their structural variability, their relatively low toxicity, and their antioxidant activities have increased the interesting flavonoids as beneficial for human health.[5]

The biological activity of flavones can be modified upon the formation of metal complexes. Flavonoid-metal complex compounds are the subject of our long-time research during which we investigated properties, composition, complex formation features, stability constants, as well as an analytical appraisal of approximately 40 complexes of flavonoids from different flavonoid subclasses with a number of metal ions or metal groups. The biological effects of flavonoid-metal complexes are showed that complexes are more effective than free flavonoids. All these data are available in various previous publications.

Flavones are medicinal natural products. Prevention and cure of disease using flavonoids as a photochemical are known due to its wide spectrum activities it attracts researchers of the world. Hence cheap easy lab feasible synthetic methodology is the demand of pharmaceutical industries. In that point of view, we optimized the Claisen-Schmidt method for the synthesis of flavones. Different molar ratios, solvent, temperature and catalysts will be examined during experiments. Also the effect of structure and substituent on the product formation and effect of groups and other on starting material will be considered.

The flavonoids or bioflavonoids are a ubiquitous group of polyphenolic substances which are present in most plants. They also occur as glycosides. Chemically, flavonoids show a fifteen-carbon skeleton (C6-C3-C6) which consists of two phenyl rings connected by a three carbon bridge. Flavonoids lead to potent antioxidant activity, the most important function of flavonoids to scavenge hydroxyl radicals, superoxide anions and lipid peroxy radicals. Multiple combinations of hydroxyl groups, sugar, oxygen and methyl groups attached to these structures create the various classes of flavonoids, flavonols, flavonones, flavones, flavon-3-ols (catechins), anthocynins and isoflavones.[6] Chalcone is an inter-mediate compound in the biosynthesis of flavonoids, which are the substances widespread in plants and with an array of biological activities. Several therapeutically interesting biological activities of certain flavonoids have been reported including antibacterial,[7] antiviral, anti-inflammatory, antiallergic, antithrombotic,[8] antimutagenic, antineoplastic, neuroprotective properties,[9] and antioxidant properties, etc.

Oxidation is one of the most important processes, it produces free radicals. In turn, these radicals start chain reactions that damage cells and cause oxidative stress. This oxidative stress in cells results in the development of a wide range of diseases like Alzheimer's disease,[10-11] Parkinson's disease,[12] the pathologies caused by diabetes[13-14] rheumatoid arthritis and neuro-degeneration in motor neuron.[15] Antioxidants flavones have been studied specifically as they terminate the chain reactions by removing free radical intermediates and inhibit other oxidation reactions.

Flavones occupy a special place in the realm of natural and synthetic organic chemistry owing to their useful biological activities such as anti-oxidant, anxiolytic, anticancer, analgesic, and antimicrobial. During the past few years various methods have been reported. Drugs having a pyrone ring system are well known for the anti-inflammatory activity. In view of above facts, it was contemplated to design and synthesis of some flavones (contains pyrone ring) derivatives from the corresponding chalcone by using NaOH and hydrogen peroxide, DMSO/I2 as an oxidizing agent.[16] The structure of flavone derivatives will be confirmed by melting point, TLC and spectral data.

Functions of Flavonoids:

1) Industrial Functions[17] :

Flavones are prototypes of phenolic polymers which are present in relatively large amount in some plant species and are of economic importance in leather tanning oil-well drilling, adhesive and other industries. Flavonoids inhibit sulphite pulping and they have been shown to be fungi static so they are important in determining the durability of timber.

2) Pharmacological Functions:

Flavones in animal have shown some interesting results such as vitamin-p-activity,[18] medicinal uses of rutin for the treatment of cold and injury in coagulation with ascorbic acid and for relieving nutritional stresses flavonoids such as myricetin, morin, kaempferol and flavonol glycosides like quercetin, rutin as well as flavon luteolin[19] have been found to possess diuretic action.

3) Enzyme Inhibitors:

Flavonoids are known to inhibit a number of enzymes such as aldose reductase,[20] lipo-oxygenase[21] and cyclooxgenase[22] etc.

4) Anti-Microbial Activity:

Antimicrobial activity has been shown by a number of flavonoids. Quercetin, Baicalin, Hesperidin, Fisetine were found to be active against *Staphylococcus aurens*. Whereas Apigenin[23-24] and Chrysin[25-27] found to be active against staphylococcus species.

5) Cardiotonic Activity:

Flavonoids have been reported to have action on the heart. The unsubstituted parent flavones exerts coronary dilatory activity and was commercially available under name "Chromocor" and its combination with rutin and iso-Quercetin was also available with the brand name "Flaco" useful in the treatment of atherosclerosis.

3-methyl-quercetin has a positive chronotropic effect on guinea pig right atrium and anti arrhythmic effect on left atrium.[28]

6) Nitrogen Fixation Precursor:

Symbiotic nitrogen fixation is an alternative of relying completely on nitrogen fixation by free-living bacteria and it is interesting that when exogenously applied Quercetin increased the number of nodules containing nitrogen-fixing bacteria in the root system of lotus corniculatous.[29]

7) Flavonoids as Light Screens:

These are quite stable and visible in the ultraviolet regions of the spectrometry play a stable role as light filters in plant growth.

8) Functions in Plants:

These compounds have high ecological importance in nature as color attractants to insects and birds as an aid to plant pollination. The flavonoids contribute to many other colors found in nature particularly yellow and orange petals. They protect plant from parasite attack and control the plant growth.

Literature Survey of Previously Related Studies:

The development of numerous newer organic chelating agents that can coordinate with metal ion has opened up a broad scope to a research scientist in the synthetic field. Literature survey reveals that several organic molecules act as chelating ligands. These organic molecules include chalcones and flavones, and many more. Since transition metal ions in particular, form many stable complexes that too with chalcones and flavones ligands, the scientists have ventured into the detailed study of transition metal flavones complexes with the help of various physicochemical techniques.

Considering this, a brief survey of literature pertaining to the present investigation on transition metal complexes of related ligands is presented below.

1.2. Brief Account of Transition Metal Complexes:

Metal Complexes of Flavone:

Omar. H. Al-Obaidi has synthesized[30] a new series of Co(II) and Mn(II) complexes with the flavones ligand. The spectroscopic method and elemental analysis verified their structures. All prepared complexes have been identified by available U.V. Visible and IR in addition, the structure of the complex was characterized by magnetic moment and molar conductance in DMSO solution. From the above studies and measurements suggest an octahedral geometry around Co(II) and Mn(II). A theoretical treatment of the formation of these complexes in the gas phase was studied. This was done using the HYPERCHEM-6-Progrm for the molecular mechanics and semi-empirical calculations (Fig.1.2).

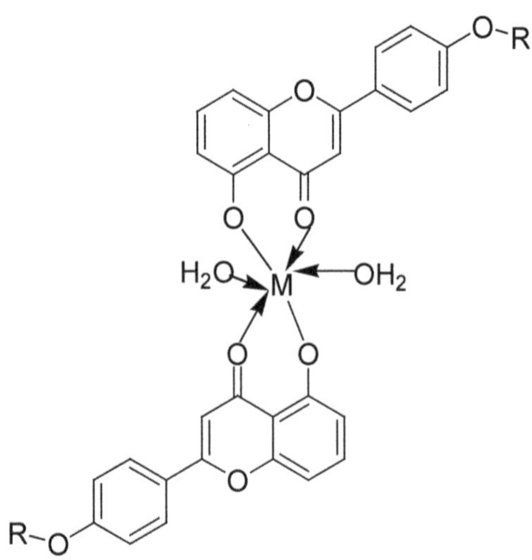

M= Co (II) and Mn (II) R= Mannose Group

Fig. 1.2: Suggested Structure of the Co(II) and Mn(II) Complexes

Yuanzhen Liu and et al.[31] has studied the systematically the effects of the number of d-electrons of the first transition metal ions such as Fe, Co,Ni, Cu and Zn on the formation and stability of metal flavonoid complexes. Based on the considerable structural information obtained through ESI-MS^n metal ions were found to form different complexes with quercetin through its 5-OH and 4-carbonyl groups. The complex stability is highly dependent on both the metallic ions and the flavonoid chelater itself. Quercetin to form the complex, in quercetin complex the stability of the complex decreased in the following order $Cu^{+2}>Ni^{+2}>Co^{+2}>Fe^{+2}>Zn^{+2}$ with flavonoid: metal stoichiometries at 1:2. The complex formed between quercetin and metal ions ($Ni^{+2}Co^{+2}Fe^{+2}Zn^{+2}$) have similar fragmentation mechanism, while Cu^{+2} displayed different fragmentation mechanism due to concurrent oxidation (Fig. 1.3).

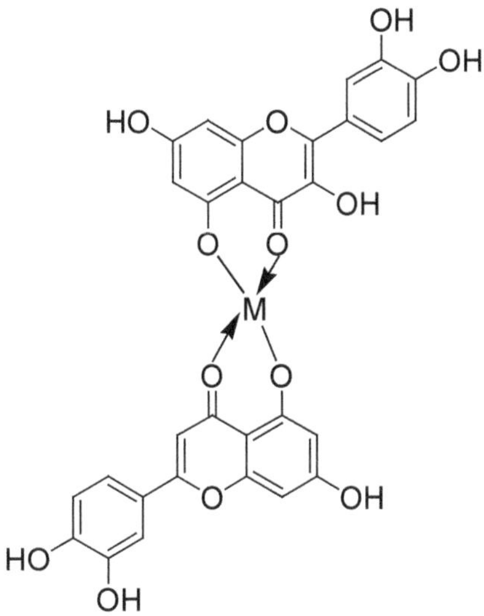

Fig. 1.3: M= Cu (II), Co (II), Ni (II), Fe (II) and Zn (II).

Safana Ahmed Farhan prepared[32] and studied the complex formation between copper (II) chloride dihydrate and morin. Structure of the complex was determined through various analytical techniques including UV-Visible, FT-IR and elemental analysis. The result of these analyses indicates that the complex has a Cu(II) ion coordinate via position 4 C=O and 5-OH of the morin and has the general structure $[CuL_2(H_2O)_2]$. The antibacterial and antifungal activities of the complex evaluated by using a wells-diffusion method and compared with the activity of free morin. This shows that, the antimicrobial activity of complxed morin has higher value as compared to the free morin. (Fig.1.4).

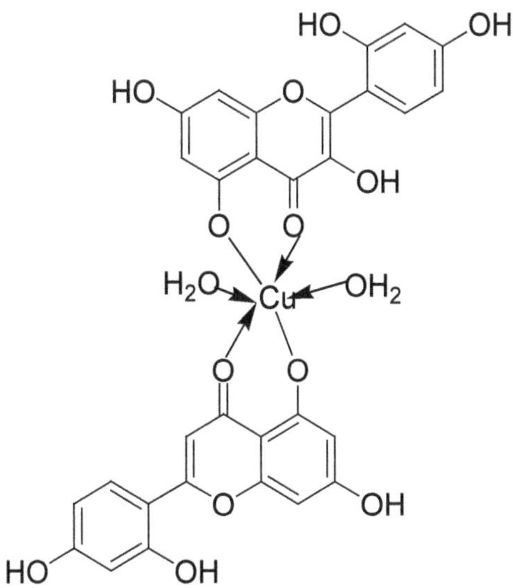

Fig. 1.4: The Possible Structure of the Cu(II)-Morin Complex.

Faiz Beg and et al.[33] have reported the effect of chrysin and its metal conjugates chrysin-copper and chrysin-Manganese on auxotrophic strains of Escherichia coli for lysine and Vitamin B12. Chrysin is reported to be an anticancer agent in a number of studies. It was complexed with Manganese for the first time to enhance its activity. The formation of a metal complex of Chrysin was confirmed by the UV-Visible, FTIR and Mass spectroscopy studies. Chrysin and its metal conjugates act as mutagenic agent in conversion of auxotrophs to prototrophs. Metal complex formation is useful to enhancing the activity of ligands (Fig.1.5).

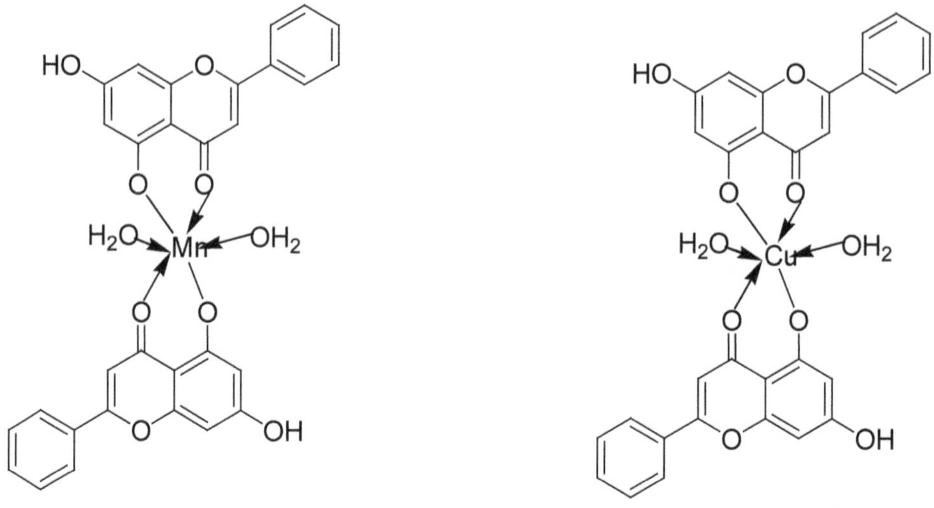

Mn-(Chrysin)$_2$ Complex Fig. 1.5. Cu-(Chrysin)$_2$ Complex

Sudhanshu S. Rout and et al.[34] synthesized new metal complexes of Co(II) with naringin and characterized by FT-IR, UV-VIS spectrophotometer. Groups of six mice were treated with cobalt narigin complex administered intraperitonealy. Different tests were used for evaluating the anti-nociceptive effect. The data obtained from characterization suggested that the coordination of Co(II) is in position 4 and 5 of the condensed ring of naringin and the increased biological activity of the synthesized complex could be due to the coordination of Co(II) in this position. The analgesic activity potentiated with CNC, in which the metal toxicity of cobalt is minimized and the efficacy, presence of two molecules enhanced, which indicate a new light of therapeutic applicability of the natural molecules (Fig. 1.6).

Qudeer K. Panhwar and et al.[35] prepared the metal complexes of rutin one of the excellent chelating agents, has been made to react with Zr^{+4} and MnO^{-4} ions to get metal complexes in methanol. The spectral properties FT-IR, UV-Visible H[1]NMR, C[13] Mass spectroscopy thermal behavior and antioxidant activity of these complexes have been studied. It has been observed that rutin forms deep yellow colored complexes of mononuclear composition having stoichiometry 1:2.

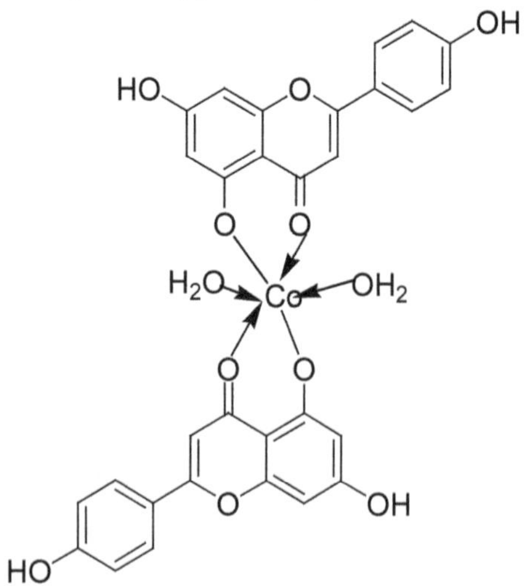

Fig. 1.6: Tentative Structure of Co (II) of Narigin Complex.

It has been further noticed that these complexes are formed at different pHs i.e. 4.0 and 8.0 suitable for Zr^{+4}

and MnO^{-4} complexes respectively. Thermograms also confirm the results of IR and support the presence of water molecules in these complexes. Relative antioxidants activity of rutin ligand and rutin complex was determined. It was found that both metal complexes have higher antioxidant activity than rutin alone. (Fig.1.7).

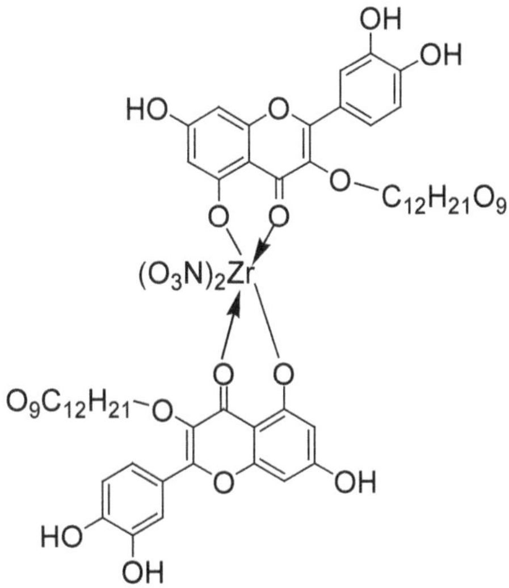

Fig. 1.7: Tentative Structure of [$Zr(C_{27}H_{28}O_{16})_2(NO3)_2].4H_2O$

J. Lachman and et al.[36] reported that the effect of abiotic stress caused by cadmium on the total flavonoid content in roots, shoots and leaf blades of spring barley variety Kompakt growth in a climate-control room for 28 days was investigated. Total flavonoid content (F) was determined spectrophotometrically with sodium nitrite, Cd content by atomic absorption spectrometry. Treatment of barley plants with Cd in nutrition solution caused the decrease of flavonoid in all parts of the plant. Statistically significant decrease of flavonoid content and increase of Cd content in all the investigated parts of the plants found (Fig. 1.8).

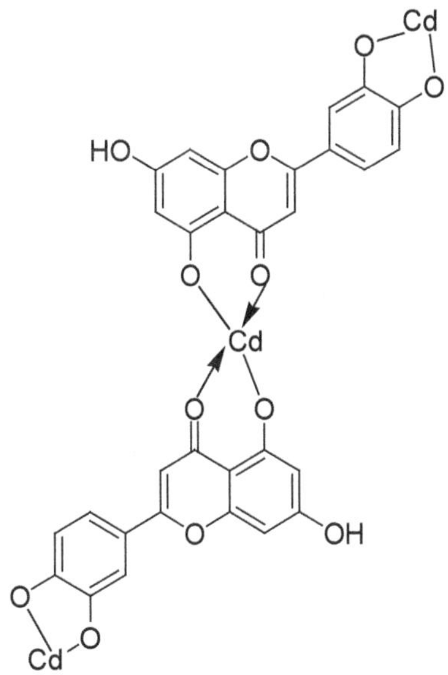

Fig. 1.8: Proposed Structure of luteolin-Cd-Complex

Venkatachalam, et al.[37] prepared a series of newer flavonols and their zinc complexes were synthesized and characterized by IR, ^{1}HNMR, and Mass spectroscopy. Their oxidation potential was determined by cyclic voltammetric method. Further they were evaluated for their cytotoxic efficacy by MTT assay using vero and MCF-7 cell lines and compound F4 and its zinc complex was found to be exhibiting cytotoxicity against MCF-7 cell lines and their IC_{50} values are comparable with quercetin. (Fig.1.9).

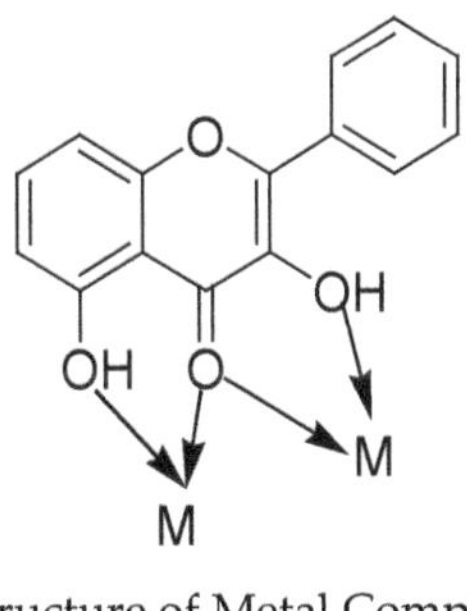

Fig. 1.9. Structure of Metal Complex $M=Zn^{+2}$

Jingliang Zhang et al.[38] synthesized a rutin-Ca-complex in methanol and characterized using elemental analysis, UV-Visible spectroscopy, IR spectroscopy and differential thermal analysis. Spectroscopic data suggested that the Metal:Ligand ratio of the complex is 2:1. The antihyperlipidemic activity of the complex was evaluated by an animal experiment of 30 Wistar rats. It was shown that the complex was much more effective antihyperlipidemic than the free flavonoid (Fig.1.10).

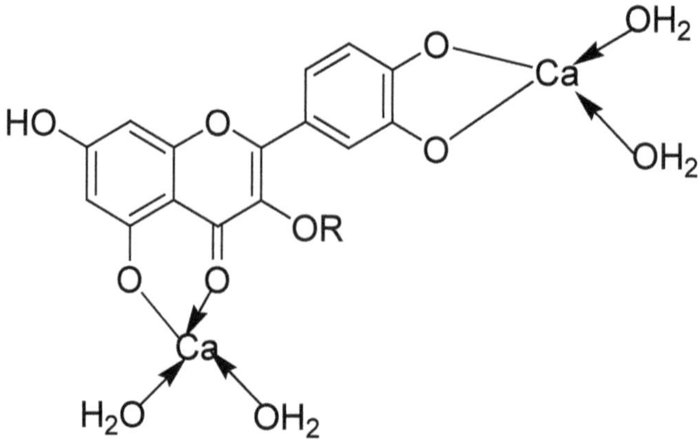

Fig. 1.10. Structure of Rutin-Ca-Complex

Valentina Ulvarosi and et al.[39] prepared a new potentially biologically active complex of oxovanadium (IV) with naturally occurring flavanone naringenin found in citrus fruits has been synthesized and characterized in a solid state by elemental analysis, molar conductance, FT-IR, UV-Visible spectra, thermal analysis and fluorescence spectra. From the experimental data, the complex was formulated as $[VO(Nrg)_2].5H_2O$ (Fig.1.11).

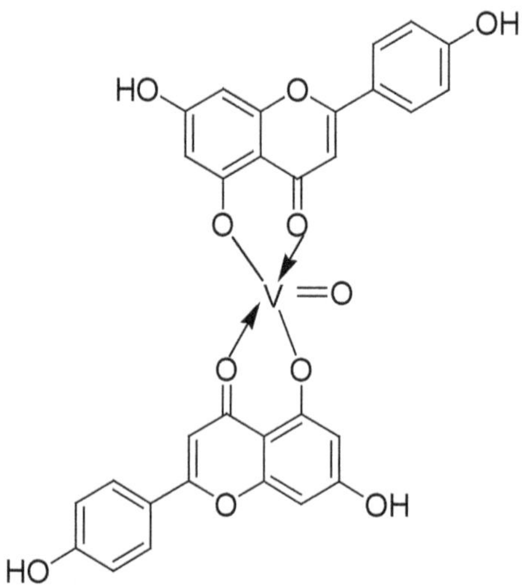

Fig. 1.11. The Possible Structure of Complex $[VO(Nrg)_2].5H_2O$

1.3. Scope of the Present Investigation:

It seems from the literature that transition metal complexes of flavones have received the least attention. On the biological front, many of the chalcones and flavones were reported to be physiologically active and find applications in the treatment of many diseases. The biological activities of these compounds have often been attributed to their ability to complexes with the metals present in the biosystem.

Considering these aspects, studies on transition metal complexes of flavones have been undertaken from our laboratory. The literature survey reveals that no work has been done on transition metal complexes of flavones. These flavones and their metal complexes have been proved to have good antifungal and antibacterial activities. Hence in continuation with previous reported studies and in view of the significant characteristics of flavones, the present work is undertaken.

The following series of flavones have been synthesized.

6) B1: 2-(furan-2-yl)-5-hydroxy-4H-chromen-4-one

7) B2: 5-hydroxy-2-(5-methylfuran-2-yl)-4H-chromen-4-one

Since the flavones and metals of higher oxidation states, both show antimicrobial activities it was thought worthwhile to see their combined effect in the form of metal complexes.

Hence for the synthesis of complexes from these, flavones we have selected different metal ions having highest oxidation states, Mn(II), Co(II), Ni(II), Cu(II) Zn(II) and Cd(II) metals ions as they are having large applicational areas.

Nomenclature, Chemical Formula and Structure of Flavone Ligands:

Sr. No.	Nomenclature and Chemical Formula	Symbol (Mol.Wt.)	Molecular Structure	Donor Atoms
6	2-(furan-2-yl)-5-hydroxy-4H-chromen-4-one Chemical Formula : $C_{13}H_8O_4$-	B1 228	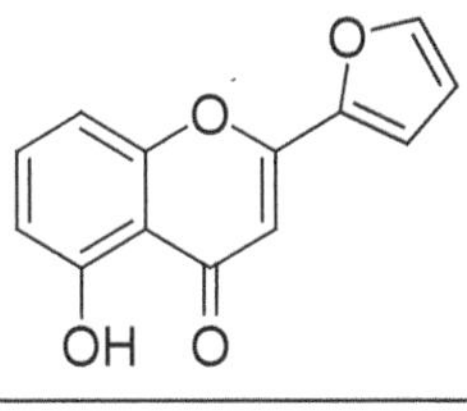	OO
7	5-hydroxy-2-(5-methylfuran-2-yl)-4H-chromen-4-one Chemical Formula : $C_{14}H_{10}O_4$	B2 242	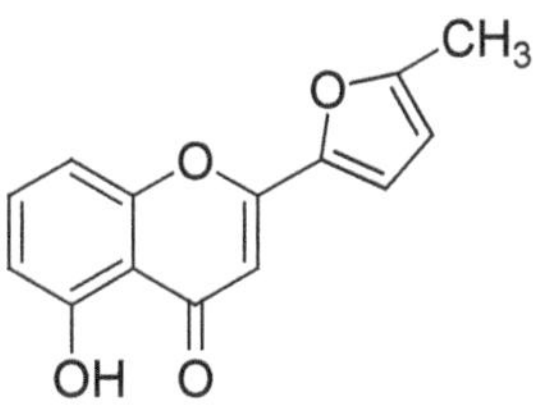	OO

CHAPTER 2

Synthesis and Characterization of Flavones and their Metal Complexes

2.1. Basic Requirements:

Apparatus:

The experimental work was carried out using borosilicate glass apparatus. All the beakers, flasks and other glass wares used for experimental work scrupulously cleaned before use by means of chromic acid cleaning agent followed by tap water and portions of deionised water. Finally, they were weighed on digital balance with 0.01 mg sensitivity.

Chemicals: All the chemicals used for experimental work were of AR grade.

a) Water and Other Solvents:

Throughout the experimental work, the glass distilled water was used. The glass distilled water was obtained by distillation of metal distilled water in the presence of crystals of potassium permanganate.

For the synthesis of Chalcone, flavones and their transition metal complexes, ethanol or methanol was used as solvents. "Anhydrous methanol " and super dry "ethanol" were obtained from commercial methanol (AR) and absolute ethanol respectively, by distillation after treatment with magnesium metal turnings and iodine. All other solvents used during the experimental work were of AR grade.

b) Reactants:

AR grade reactants such as 2,6-dihydroxy acetophenone and 2-Furaldehyde, 5-methyl-Furaldehyde, used for the synthesis of ligands. Metal salts are used for the synthesis of metal complexes. The purity of chemicals was checked by routine tests like melting point, thin layer chromatography. The starting material recrystallized from ethanol. Purity of starting material was checked by thin layer chromatography and melting point.

2.2. Synthesis of Flavones:

Synthesis of A_1 Ligand:

A mixture of 2,6-dihydroxy acetophenone (0.01 mol) and 2-furaldehyde (0.01 mol) are dissolved in ethanol (20 mL) and then solution of potassium hydroxide 10 mL (15%) were added to it. The mixture was stirred for overnight. The progress of the reaction was monitored by TLC. It was then poured on ice cold water and acidified with dilute HCl. The coffee brown solid was precipitates, filtered and washed with water and recrystallized from ethanol.

Synthesis of A_2 Ligand:

The mixture of 2,6-dihydroxy acetophenone (0.01 mol) and 5-methyl-Furaldehyde (0.01 mol) is dissolved in ethanol (20 mL) and then sodium hydroxide 10 mL (40%) was added to it. The mixture was stirred for overnight till brown color ppt was observed. The progress of the reaction was monitored by TLC, from thin layer chromatography the completion of the reaction is observed. After completion of the reaction, the contents were poured into ice cold water and then acidified by dil. HCl. The solid obtained was filtered and crude product was recrystallized from ethanol to give the chalcone.[40]

Synthesis of $B_1$and B_2 Ligand:

Take A_1 /A_2 ligand (0.01 mol) and then it was dissolved in 20 ml DMSO, to this catalytic quantity of iodine was added. Contents were refluxed for one hour, the progress of the reaction was monitored by TLC and the reaction mixture was left overnight. It was then poured on ice cold water, the separated solid was filtered washed with cold water followed by a dilute sodium-thiosulphate solution. The product was crystallized from ethanol.[41]

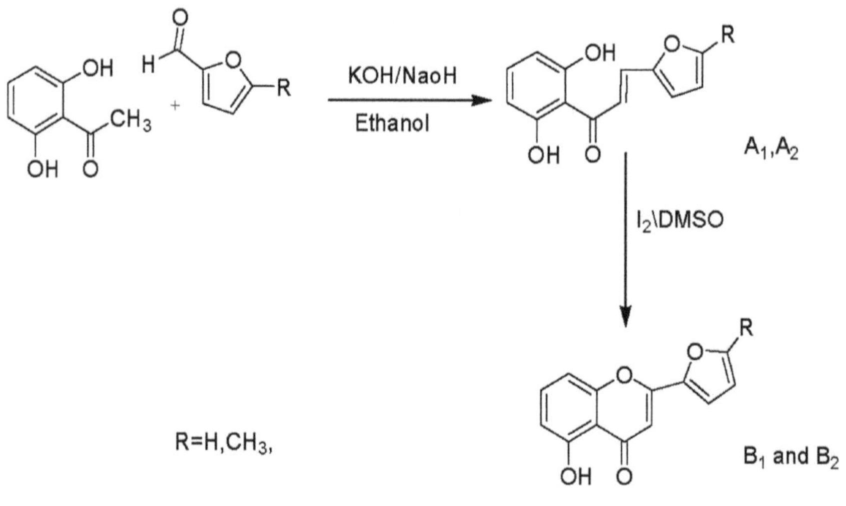

2.3. Characterization of Flavones:

All synthesized ligands are stable to air and moisture, soluble in DMF, DMSO and insoluble in ethanol, methanol, chloroform, ether and water. The structural features of the ligands are elucidated with the help of Elemental Analysis, Infrared, ^{1}H-NMR and Mass spectral techniques.

Elemental Analysis, Color and Melting Points of Ligands:

The elemental analyses (C,H,N,O) were performed on Perkin Elmer analyzer.

The color, melting point, molecular weight and percentage of carbon, hydrogen, nitrogen, oxygen and bromine found and calculated theoretically are given in Table 2.1

Table 2.1 : Elemental Analysis Data:

L	Mol. Formula.	Color	Mol. Wt.	M.P °c	Found (Calculated) %					
					C	H	O	N	S	Br
B_1	$[C_{13}H_8O_4]$	Coffee brown	228	272	68.40 (68.42)	3.49 (3.53)	28.00 (28.04)	---	---	---
B_2	$[C_{14}H_{10}O_4]$	Yellowish brown	242	220	69.46 (69.42)	4.19 (4.16)	26.47 (26.42)	---	---	---

Spectral Study of Flavones:

Spectroscopy is the study of the interaction of electromagnetic radiation with matter. The ultraviolet-visible (UV-VIS), infrared (IR), nuclear magnetic resonance (NMR) and mass spectroscopy are the important spectroscopic techniques for the structure determination of organic ligands.

Electronic Absorption Spectral Study of Flavones (UV-Visible):

Absorption in the ultraviolet and visible region of the electromagnetic spectrum depends on the electronic structure of molecules. In practice, the ultraviolet spectroscopy is mainly used to measure the multiple bonds and aromatic conjugation within a molecule. The energy absorbed in ultraviolet region produces changes in the electronic energy of molecule resulting in the transition of valance electrons. These transitions consist of the excitation of electrons from an occupied molecular orbital to higher energy unoccupied molecular orbital. The electronic transitions that are involved in ultraviolet and near visible regions are of the types σ-σ^*, n-σ^*, n-π and π-π^*. But among these transitions, significant transition is π-π^*. The energy required for different transitions is in the following order n-π^* < π-π^* < n-σ^* < σ-σ^*.

The spectroscopic study in UV/Vis region for convenience, electronic transitions are assigned by Buraway as R, K, B and E bands.

Type of band	Electronic transition	ε_{max}
R	n-π^*	<100
K	π-π^*	>10,000
B	π-π^*	~1,000 to 5,000
E	π-π^*	2,000 to 14,000

The solutions of known concentration (10^{-3}M) of flavone ligands were prepared in alcohol and their UV/Vis spectra were recorded on SL159, single beam UV-VIS spectrophotometer in the range 200-800 nm. The ligand solution spectra were recorded by filling a reference cell with pure solvent.

The electronic absorption spectra of flavones are presented in Fig.2.1. to 2.2. The electronic absorption peak positions were used to assign the λ_{max} values to prominent chromophores, such as >C=O, -OH group present in the ligands.

Generally, the flavones and flavonols exhibit absorption in 320-380 nm regions (band I) and 240-270 nm regions (band II).

Table 2.2: UV/ Vis Spectral Data of Flavones

Sr. no.	Ligands	Symbol	λ_{max} for > C=O in nm	λ_{max} for -OH in nm
6	2-(furan-2-yl)-5-hydroxy-4H-chromen-4-one	B_1	360	245
7	5-hydroxy-2-(5-methylfuran-2-yl)-4H-chromen-4-one	B_2	339	258

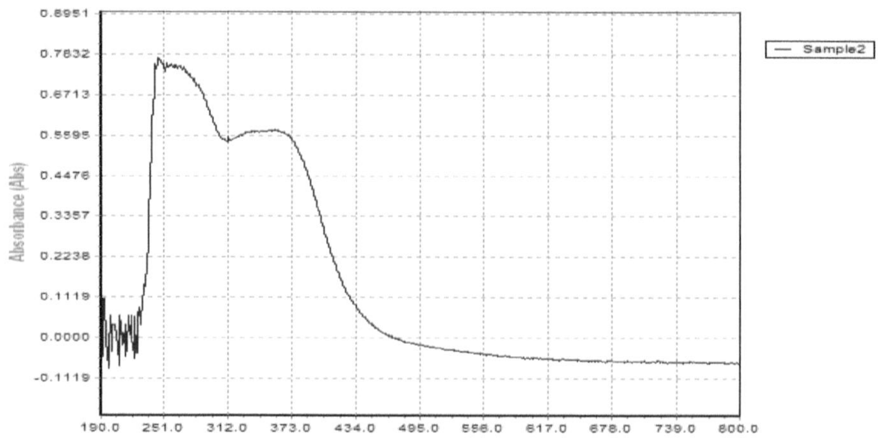

Fig. 2.1: Electronic Absorption Spectrum of Flavones B_1

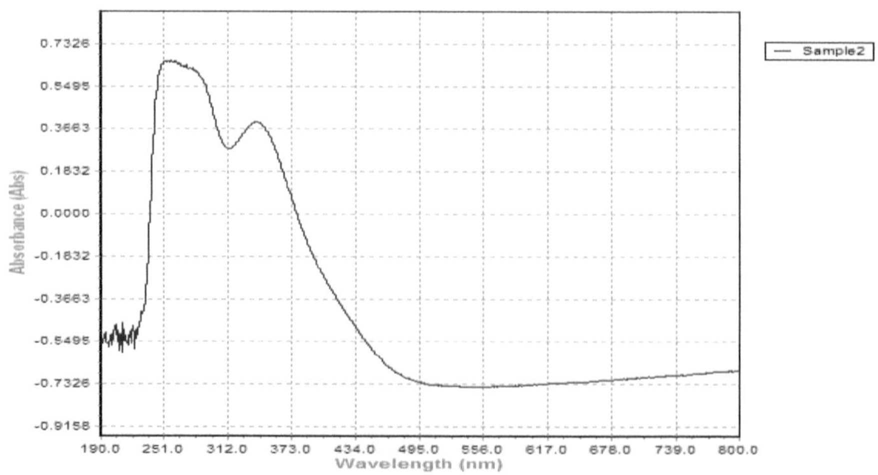

Fig. 2.2: Electronic Absorption Spectrum of Flavones B_2

^{1}H NMR Spectral Study of Flavones:

Nuclear Magnetic Resonance Spectroscopy is possible due to the absorption of energy at particular frequencies by atomic nuclei, characterized by a property, termed spin and this gives rise to a magnetic moment associated with that nuclei. The present discussion is related to the proton nucleus having spin moment 1/2 can resonate to give absorption at a characteristic frequency. The frequency at which proton absorbs energy depends on electronic environment present around that proton. In the analysis of organic molecule, ^{1}H NMR spectra play a very important role. It is the most valuable technique in structural investigation.

As any spinning charge body generates a magnetic field, the proton behaves like a tiny spinning bar magnet, orients itself under the influence of external magnetic field. The frequency of the group of nuclei can be measured with respect to some reference like tetramethylsilane (TMS). The chemical shift of proton is obtained by

$$\delta = \frac{\Delta \upsilon}{\upsilon_0}$$

Where ,

δ = chemical shift in ppm

Δ υ=Absorption frequency difference in hertz (Hz)

υ_o=Operating frequency of the instrument in mega heartz (MHz)

The chemical shift (δ) of proton depends upon the factors such as electronegativity, electron density, which causes particular nuclei to appear at different chemical shift (δ). Greater the shielding effect, lower the chemical shift and opposite of this character is deshielding effect.

The ^{1}H NMR spectra are recorded on Bruker AVANCE II 400 MHz Spectrophotometer in DMSO solvent using TMS as an internal standard at SAIF, Chandigarh, Punjab.

Discussion of ^{1}H NMR Spectral Data of Flavones:

The formation of flavones and their structures are confirmed by ^{1}H NMR spectral study of ligands. The result obtained was used to interpret the proton environment and the number of protons present in the sample. The NMR spectra of ligands are presented in Fig. 2.3 to 2.4 and the characterizations of particular protons are presented in Table 2.3 to 2.4.

In flavones, the ^{1}H NMR spectrum appearance of a signal at δ-6.9-6.67 singlets (1H of pyran ring) supported the flavones structures.[41]

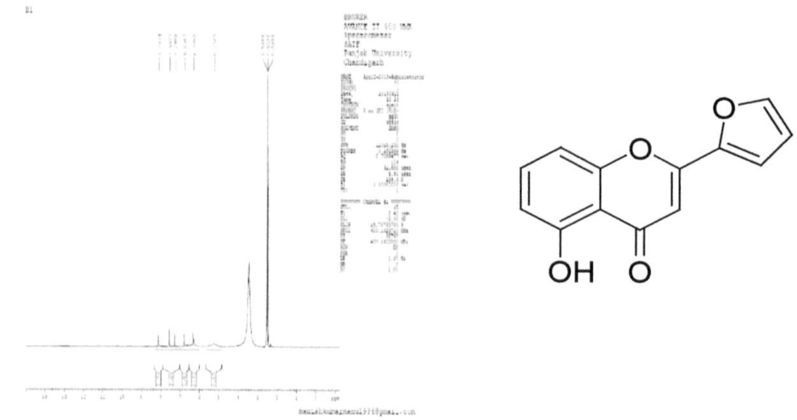

Fig. 2.3. [1]H Nuclear Magnetic Resonance Spectrum of Flavones B_1

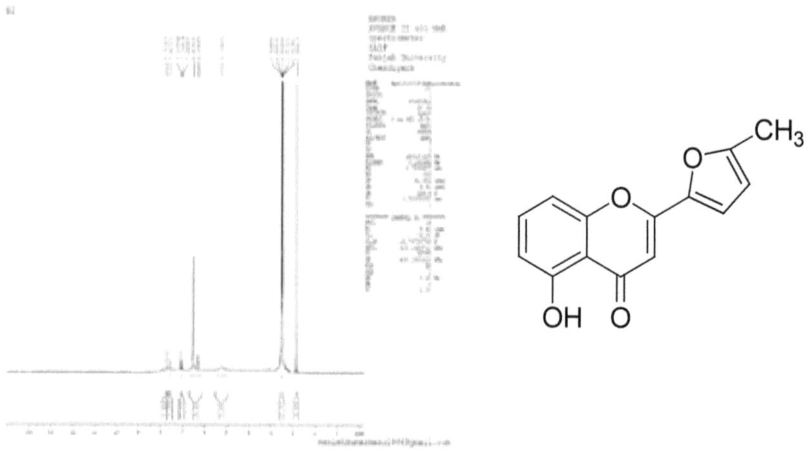

Fig. 2.4: [1]H Nuclear Magnetic Resonance Spectrum of FlavonesB_2

2-(furan-2-yl)-5-hydroxy-4H-chromen-4-one

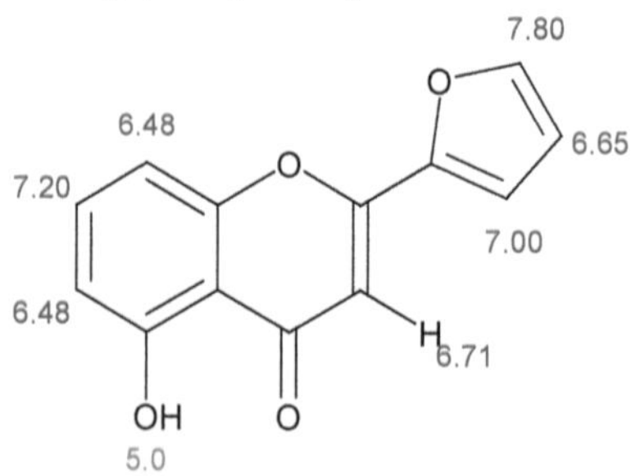

Chemical Formula:$C_{13}H_8O_4$

Fig. 2.5: [1]H Nuclear Magnetic Resonance Spectral Data of Flavones B_1

Table 2.3 Flavone B_1

Chemical Shift (δ) ppm	Number of Protons	Multiplicity (Splitting)	Assignment
6.48-7.20	3H	m	Aromatic protons
5.0	1H	s	-OH group present on the aromatic benzene ring
6.71	1H	S	α-H on-unsaturated carbonyl system
6.65-7.80	3H	m	Protons on furan ring

5-hydroxy-2-(5-methylfuran-2-yl)-4H-chromen-4-one

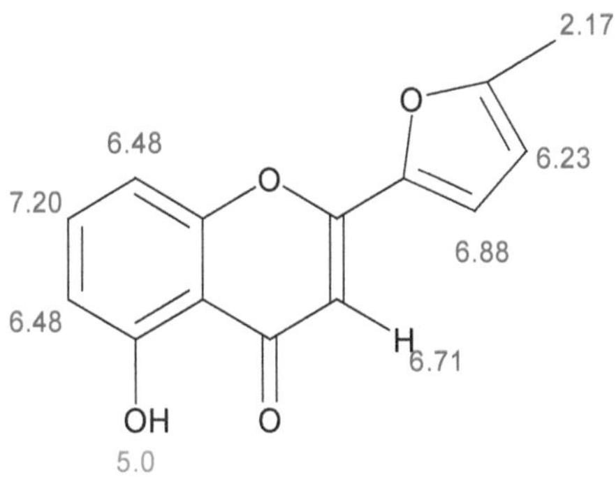

Chemical Formula:$C_{14}H_{10}O_4$

Fig. 2.6. [1]H Nuclear Magnetic Resonance Spectral Data of Flavone B_2

Table 2.4 Flavone B_2

Chemical Shift (δ) ppm	Number of Protons	Multiplicity (Splitting)	Assignment
6.48-7.20	3H	m	Aromatic protons
5.0	1H	s	-OH group present on the aromatic benzene ring
6.71	1H	s	α-H on-unsaturated carbonyl system
6.23-6.88	2H	d	Protons on furan ring
2.17	3H	m	Methyl protons attached to furan ring

Mass Spectra of Flavone:

Mass spectroscopy is the most accurate technique for the determination of the molecular weight of the compound. In this technique matter is bombarded with highly energetic electrons. Then matter absorbs electrons or ejects electrons, from itself. When it ejects electrons charged species are formed.

$$M - e^- \longrightarrow \overset{\bullet +}{M} + 2e^-$$ Leaves behind a radical cation.

$$M + e^- \longrightarrow M^{\bullet -}$$ ·leaves behind an anionic radical.

Anionic and cationic radicals are detected by the detector. Detector never detects radical and neutral species. The mass spectrum is a plot representing the m/e values of various ions against their relative percent intensity. The highest intense peak in the spectrum is called the base peak. The intensity of other peaks is shown relative to the base peak. The peak at extreme right corresponds to the molecular weight of the original molecule. The molecular ion is called parent ion and usually denoted as $[M]^+$ ion. Mass spectroscopy also separates the isotopes.[42]

The mass spectra of ligands were recorded on Waters, Q-TOF Micro Mass (LC-MS) at SAIF, Chandigarh, Punjab.

The molecular masses of ligands and theoretically calculated are presented in Table 2.5 Mass spectra are presented in Fig. 2.7-2.8.

Table 2.5

Ligands	**B_1**	**B_2**
$[M^+]$ Molecular Mass Found (Calculated)	229	242

In the mass spectrum of the ligands, molecular ion peak is observed m/z and calculated m/z corresponding to $[M^+]$, peaks are in good agreement with their proposed ligand structures.

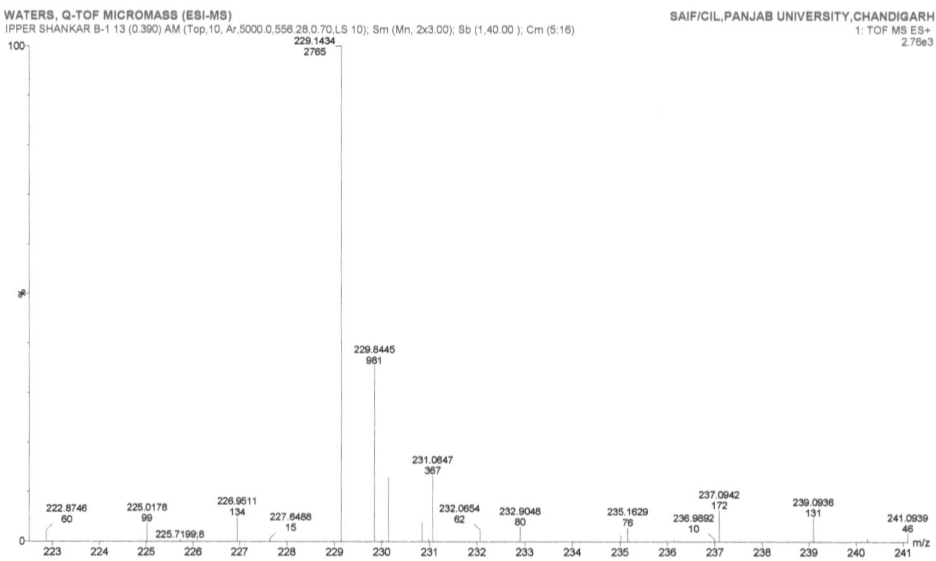

Fig.2.7: Mass spectra of Flavone B_1

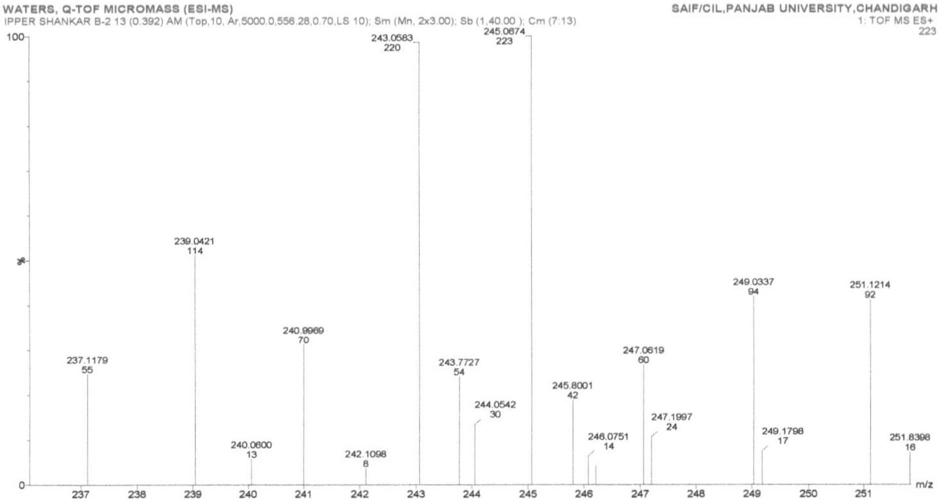

Fig.2.8: Mass spectra of Flavone B_2

Table 2.6: Nomenclature, Chemical Formula and Structure of Flavone Ligands

Sr. No.	Nomenclature and Chemical Formula	Symbol (Mol.Wt.)	Molecular Structure	Donor Atoms
6	2-(furan-2-yl)-5-hydroxy-4H-chromen-4-one Chemical Formula : $C_{13}H_8O_4$	B1 228	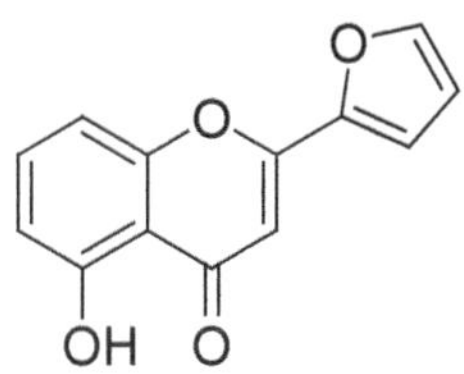	OO
7	5-hydroxy-2-(5-methylfuran-2-yl)-4H-chromen-4-one Chemical Formula : $C_{14}H_{10}O_4$	B2 242	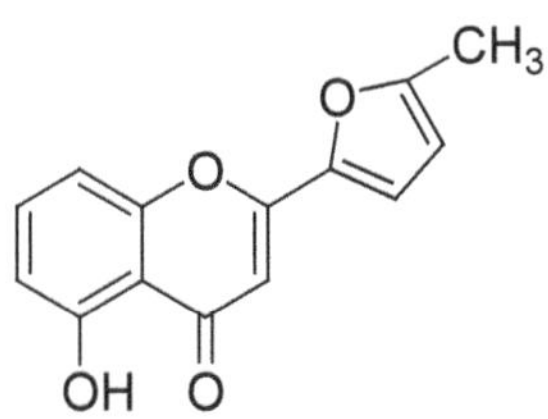	OO

2.4. Synthesis of Metal Complexes:

The metal complexes are synthesized by various methods. In the present work, the transition metal complexes of flavone are synthesized by stirring the methanolic solution of all ligands and Metal Acetates of Mn(II), Metal Sulphates of Cu(II), Metal Nitrates of Co(II), Ni(II) Zn(II) & Cd(II) in the 2:1 molar ratio. The slightly alkaline condition maintained by adding 10% alcoholic ammonia solution.

Procedure:

The solid 0.02 mole of ligand (flavone) was taken in a round-bottomed flask containing 30 mL of anhydrous methanolic solution of the ligand was used and boiled for a few minutes. A hot solution of 0.01 mole of the metal salt in 20 ml of methanol was added drop wise to the solution of the ligand. To this reaction mixture, 10% alcoholic ammonia was added up to slightly alkaline pH. The complexes of different metals were precipitated at different pH range. This pH range was definite for a given complex and found to be characteristic of that complex. The contents were stirred on magnetic stirrer for one hour. The solid metal complex separated out and washed with methanol three to four times. Dried in vacuum desiccators over anhydrous granular calcium chloride.[43]

The metal salts used for the synthesis of metal complexes were Mn(II) Acetate, Cu(II) Sulphate, and Nitrates of Co(II), Ni(II), Zn(II) and Cd(II).

The melting point/decomposition temperatures of the complexes were determined by Thiele's melting apparatus. The pH range, observed melting points and colors of metal complexes are tabulated in Table 2.7 to 2.12.

Table 2.7: Physical Parameters of Manganese (II) Complexes.

Sr. No.	Manganese (II) Complexes of	pH range of precipitation	Color	**Melting point/ Decomposition Temp. °C**
1.	B_1	7.5-8.5	Blackish brown	>300
2.	B_2	7.5-8.5	Brown	>300

Table 2.8: Physical Parameters of Cobalt (II) Complexes.

Sr. No.	**Cobalt (II) Complexes of**	**pH range of precipitation**	**Color**	**Melting point /Decomposition Temp. °C**
1.	B_1	7.5-8.5	Brown	>300
2.	B_2	7.5-8.5	Greenish	>300

Table 2.9 : Physical Parameters of Nickel (II) Complexes.

Sr. No.	**Nickel (II) Complexes of**	**pH range of precipitation**	**Color**	**Melting point / Decomposition Temp. °C**
1.	B_1	7.5-8.5	Blackish brown	>300
2.	B_2	7.5-8.5	Greenish	>300

Table 2.10 : Physical Parameters of Copper (II) Complexes.

Sr. No.	**Copper (II) Complexes of**	**pH range of precipitation**	**Color**	**Melting point / Decomposition Temp. °C**
1.	B_1	7.0-7.5	Reddish	>280
2.	B_2	7.0-7.5	Coffee brown	>260

Table 2.11: Physical Parameters of Zinc (II) Complexes.

Sr. No.	**Zinc (II) Complexes of**	**pH range of precipitation**	**Color**	**Melting point / Decomposition Temp. °C**
1.	B_1	7.0-7.5	Brown	290
2.	B_2	7.0-7.5	Reddish brown	>280

Table 2.12 :Physical Parameters of Cadmium (II) Complexes.

Sr. No.	Cadmium (II) Complexes of	pH range of precipitation	Color	Melting point / Decomposition Temp. °C
1.	B_1	7.5-8.5	Brown	>300
2.	B_2	7.5-8.5	Reddish brown	>300

2.5. Characterization of Metal Complexes:

All the metal complexes of flavones are stable to air and moisture. The solubility of these complexes was examined in different solvents like water, polar and non-polar organic solvents. All the chalcone and flavone transition metal complexes are insoluble in water. Majority of complexes are sparingly soluble in chloroform, ethanol and dioxane and easily soluble in dimethyl sulphoxide (DMSO) and dimethylformamide (DMF).

The synthesized metal complexes are characterized by elemental analysis and mass spectroscopy.

Table 2.13: Analytical Data of Mn(II) Complexes.

Mn (II) Complexes of Ligands	Chemical Formula	Mol. Wt.	Elemental analysis :% Found (Calculated)						
			C	H	N	O	S	X (Br)	M
B_1	$[C_{26}H_{18}O_{10}Mn]$	545	57.30 (57.26)	3.32 (3.33)	-	29.30 (29.34)	-	-	10.09 (10.07)
B_2	$[C_{28}H_{22}O_{10}Mn]$	573	58.50 (58.65)	3.89 (3.87)	-	27.96 (27.90)	-	-	9.67 (9.58)

Table 2.14: Analytical Data of Co(II) Complexes.

Co (II) Complexes of Ligands	Chemical Formula	Mol. Wt.	Elemental analysis :% Found (Calculated)						
			C	H	N	O	S	X (Br)	M
B_1	$[C_{26}H_{18}O_{10}Co]$	549	56.81 (56.84)	3.25 (3.30)	-	29.09 (29.12)	-	-	10.61 (10.73)
B_2	$[C_{28}H_{22}O_{10}Co]$	577	58.29 (58.24)	3.88 (3.84)	-	27.77 (27.71)	-	-	10.30 (10.20)

Table 2.15: Analytical Data of Ni(II) Complexes.

Ni (II) Complexes of Ligands	Chemical Formula	Mol. Wt.	Elemental analysis :% Found (Calculated)						
			C	H	N	O	S	X (Br)	M
B_1	$[C_{26}H_{18}O_{10}Ni]$	549	56.76 (56.87)	3.28 (3.30)		29.03 (29.13)	-	-	10.58 (10.68)
B_2	$[C_{28}H_{22}O_{10}Ni]$	577	58.31 (58.26)	3.92 (3.84)		27.82 (27.72)	-	-	10.28 (10.16)

Table 2.16: Analytical Data of Cu(II) Complexes.

Cu (II) Complexes of Ligands	Chemical Formula	Mol. Wt.	Elemental analysis :% Found (Calculated)						
			C	H	N	O	S	X (Br)	M
B_1	$[C_{26}H_{18}O_{10}Cu]$	553	56.32 (56.37)	3.31 (3.28)	-	28.81 (28.88)	-	-	11.33 (11.47)
B_2	$[C_{28}H_{22}O_{10}Cu]$	582	57.85 (57.78)	3.89 (3.81)	-	27.55 (27.48)	-	-	11.00 (10.91)

Table 2.17: Analytical Data of Zn(II) Complexes.

Zn (II) Complexes of Ligands	Chemical Formula	Mol. Wt.	Elemental analysis :% Found (Calculated)						
			C	H	N	O	S	X (Br)	M
B_1	$[C_{26}H_{14}O_8Zn]$	519	60.12 (56.18)	2.71 (3.26)	-	24.62 (28.79)	-	-	12.57 (11.76)
B_2	$[C_{28}H_{18}O_8Zn]$	547	61.38 (57.60)	3.31 (3.80)	-	23.36 (27.40)	-	-	11.93 (11.20)

Table 2.18: Analytical Data of Cd(II) Complexes.

Cd (II) Complexes of Ligands	Chemical Formula	Mol. Wt.	Elemental analysis :% Found (Calculated)						
			C	H	N	O	S	X (Br)	M
B_1	$[C_{26}H_{18}O_{10}Cd]$	602	51.86 (51.80)	2.95 (3.00)		26.60 (26.54)	-	-	18.69 (18.65)
B_2	$[C_{28}H_{22}O_{10}Cd]$	630	57.65 (57.60)	3.57 (3.50)	-	25.42 (25.36)	-	-	17.89 (17.81)

Mass Spectral Study of Metal Complexes:

The mass spectra of complexes were recorded on Waters, Q-TOF Micro Mass (LC-MS) at SAIF, Chandigarh, Punjab. Mass spectroscopy gives the molecular mass of the compound. The mass spectra of representative metal complexes were recorded.

The mass spectra of the complexes support their proposed structure.[44] All complexes show $[M^+]$ molecular ion peak which exactly matches with their calculated masses. In all of these complexes, 100% abundance base peak has been shown with loss of one or two enolic protons. However, in case some of complexes, M^+ ion peak appear less by one, two or three mass units, than their molecular weights.

The mass spectra of complexes indicate that there is a formation of 1:2 metal-ligand stiochitometry.[45] The presence of water molecules observed since the molecular ion peaks of complexes exactly match with the mass of 1:2 metal ligand proportion.

The molecular mass of metal complexes found by mass spectra and theoretical calculated are presented in Table 2.19. The underlined molecular weight in table corresponds to base peak or highest intensity peak in mass spectra of complexes. The mass spectra of representative metal complexes are shown in Fig. 2.9.

Table 2.19: Mass Spectral Study of Metal Complexes

Sr. No.	Name of Complex	Calculated Mass of the Complexes	Found Mass of the Complexes	Fragmentation Peaks.
4	B_2Cu	582	610	301,338,413,437,536

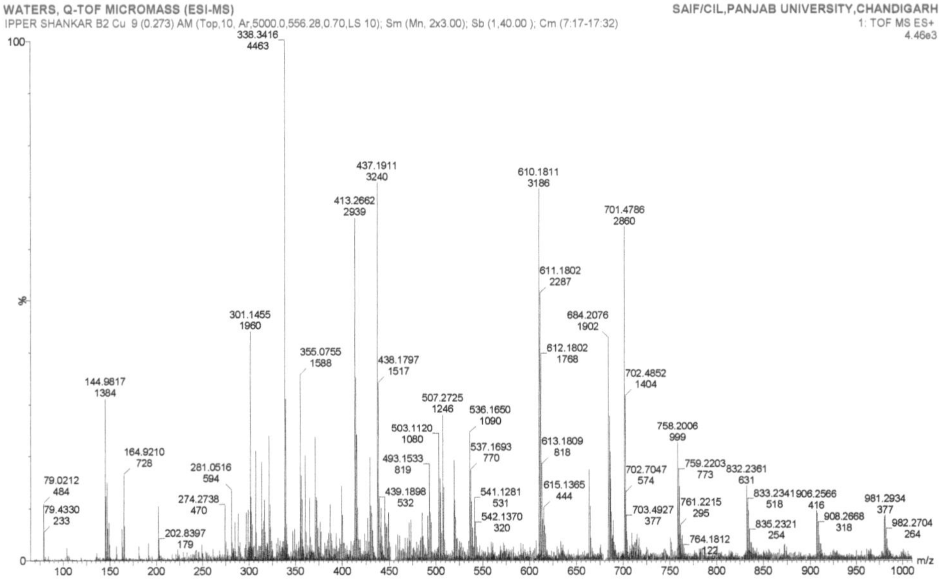

Fig.2.9: Mass spectrum of B_2Cu Complex

CHAPTER 3

Experimental Techniques

3.1. Magnetic Susceptibility:

A Magnetochemistry deals mainly with the determination of magnetic properties, has great importance in the study of transition metal complexes. The measurement of magnetic properties has contributed much towards understanding and characterization of the compounds with regard to the type of magnetism, oxidation state, stereochemistry and nature of bonding in metal complexes. The most useful parameter that is a number of unpaired electrons and their distribution in orbitals around the central metal atom determined by magnetic susceptibility. One can calculate magnetic moment (μ), the number of unpaired electrons and electronic configuration on central metal ion which decides the geometry of the metal complexes. Magnetochemistry shows whether inner or outer d-orbitals are used and whether d-electrons are paired or unpaired in the complexes. These usually distinguish the high spin and low spin complexes.[46-47]

The magnetic susceptibility indicates whether a material is attracted into or repelled out of a magnetic field. Paramagnetic materials align with the field, so are attracted to the magnetic field. Diamagnetic materials are anti-aligned, so are pushed away from magnetic fields. Secondly, on top of the applied field, the magnetic moment of the material adds its own magnetic field, causing the field lines to concentrate in paramagnetism or be excluded in diamagnetism.[48]

There are different kinds of magnetic effects shown by every molecule. The induced diamagnetic field which results in a negative susceptibility of the diamagnetic material from -10^{-6} to -10^{-5} CGC units. Paramagnetism arises in substances containing unpaired electrons and their magnetic field is parallel to the lines of force of external field. The atomic and molar susceptibilities of paramagnetic substances are of the order of 10^{-5} to 10^{-3} CGS units.[49]

The diamagnetic susceptibility is independent of temperature while paramagnetic susceptibility varies inversely with the absolute temperature. The paramagnetic effect is 10 to 1000 times larger than the diamagnetic effect.[50]

Magnetic susceptibility is a dimensionless proportionality constant that indicates the degree of magnetization of a material in response to an applied magnetic field. A related term is a magnetizability, the proportion between magnetic moment and magnetic flux density.[51]

Experimental Method:

The determination of magnetic susceptibility involves measurement of force exerted by a homogeneous magnetic field on the sample under study. If the force acting on the sample under investigation is known, the magnetic susceptibility of the compound can be determined and from this value, magnetic moment can be calculated.

Gouy's method was used for the measurements of magnetic susceptibility of the compounds at room temperature. Gouy's balance consists of an electromagnet with a suitable power supply and a single pan semi microbalance, (Mettler-Zurich, Swiss-make-H-16 GD with maximum capacity 80 gm and precision of ±0.001mg.) The electromagnet consists of cylindrical poles, 15cm in diameter, tapering at the end at an angle of 45° and ending in circular faces 3.5 cm in diameter, a gap of 4 cm is maintained between the two poles. The magnetic assembly having an electromagnet with capacity up to 7000 oersted was placed around the tube, with a current of 6 amperes fed by a suitable power supply.

The specimen tube made up of Pyrex glass, 14 cm in length and 3.5 mm in diameter with Teflon stopper is suspended by a clean silver chain from the centre of the pan of the balance in such a way that its lower end located at the centre of the pole gap. The empty, clean and dry tube weighed successively when current is off and on with 4 ampere and 6 ampere. The field strength of the magnet and position of the tube with respect to poles remain unchanged during every measurement. The mean of three readings gives the weight of the empty tube (W_g).

Calibration of the Specimen Tube:

The specimen tube was calibrated using the calibrant, Mercury tetrathiocynato cobalt (II) which was prepared by the standard method given in the literature.[2] The measurement of force developed on the specimen tube was recorded with two independent fillings of the calibrant, Mercury tetrathiocynato cobalt (II). The χ_g value of this standard substance was taken as 16.44×10^{-6} CGS units. The tube constant β was determined at room temperature. This value was used for the calculation of magnetic susceptibility of metal complexes.

A typical data of the calibrant is presented below.

Weight of empty tube = 11.90275 gm.

Force on the empty tube with applied magnetic field 4 ampere = 0.004 gm.

Weight of Hg [Co(NCS)4] (gm)	Temperature K°	Force on the sample (gm)	$\beta \times 10^6$ CGS
0.1870	303	0.004	165.67

Sources of Error:

The main source of error in the Gouy's method for solids arises from the improper packing of the sample. This can be reduced by repeating measurements on homogeneously repacked samples until relatively constant values of gram susceptibility (χ_g) are obtained. Agreement within 1% to be considered good.

Measurement of Magnetic Susceptibilities of Metal Complexes:

The finely powdered metal complex was filled in the experimental tube up to the reference mark by taping the tube on the hard surface. The precaution was taken to ensure tight and uniform packing without any air gap in the sample tube. The Teflon stopper was tightly fixed on the tube and suspended between the magnetic poles with the help of silver chain suspension from the pan of the balance described earlier. From experimental data the magnetic susceptibility (χ_g) of metal complexes were calculated by using the following expression.

$$\chi g = \frac{\beta \Delta w}{W}$$

Where,

χ_g = susceptibility per gram of the substance.

β = constant, characteristic of the tube.

Δw = force exerted on the substance.

W = weight of the substance

The molar susceptibility (χ_M) of complexes were then calculated by the equation

$\chi_M = \chi_g$ x Molecular weight.

From the values of χ_M, the magnetic moment (μ_{eff}) of complexes were calculated by the equation

$\mu_{eff} = 2.83\ (\chi_M T)^{1/2}$ B.M.

Where,

$\mu_{eff.}$ = Magnetic moment of the metal ion in Bohr Magneton (B.M.)

$\chi_A = \chi_{corr}$ = magnetic susceptibilities obtained by subtracting the molar susceptibility of ligand molecule computed by using Pascal's constants.

i.e. $\chi_{corr} = \chi_M - \chi_{diamagnetic}^{(L)}$

Where,

$\chi_{diamagnetic}^{(L)} = \sum \chi_{(atomic\ correction)}$

$T = \sum \chi$ (multiple bond)

T = temperature in Kelvin scale.

The reproducibility of magnetic susceptibility measurements of the complexes was within ± 2 %. The diamagnetic susceptibilities ($\chi_{diamagnetic}^{(L)}$) .

3.2. Solution Conductivity:

Conductivity is a measure of the ability of a solution to carry electric current. The solution conductivity depends on the number of charges on the ions, which are formed by dissociation of ionic material. The solution of an electrolyte conducts electric current by migration of ions under the influence of the electric field. Therefore by measuring the solution conductivity, one can decide the electrolytic or non-electrolytic nature of metal complexes, which helps to ascertain its ionic and covalent nature.

The molar conductivity μ_v of a solution of solute is a measure of the number and rate of migration of anions and cations in one mole of the solute. Complexes with that of known simple ionic materials, the total number of charges on the species formed, the dissolved complexes can be eluted. This decides the composition of complexes by cryoscopic measurements.[52]

Measurement of Solution Conductivity:

The solution conductivities of 10^{-3} M solution of all metal complexes in DMSO were measured on *"EQUIPTRONICS"* digital conductivity meter EQ-660 with 20 μΩ to 200 μΩ at 298K temperature. As the conductivity of solution rises by 2 percent per degree, the instrument and conductivity cell was calibrated using 0.005M KCl solution at room temperature. The observed conductance (x) of solution was measured and specific conductivity K_v was calculated by multiplying the cell constant with observed conductance. Finally, from specific conductivity and molarities (M) of the solution, the molar conductivity μ_v was calculated by the equation.[53]

$\mu_v = Kv\_x\ V_M$

$$\mu_v = \frac{K_v \times 1000}{M}$$

Where,

M is molarities of the solution of solute and has unit $ohm^{-1}cm^{2}mol^{-1}$. The molar conductance values were interpreted with the help of literature data.[54-55]

In general the more ions present in a solution the greater the conductivity, however, not all additions to aqueous solutions reliably form ions e.g. sugar and alcohol.

Further, conductivity only increases with concentration up to a maximum value, after which, the conductivity may actually decrease with increasing concentration.[56]

3.3. Electronic Absorption Spectra:

Ultraviolet and visible spectroscopy is primarily used to measure the multiple or aromatic conjugation within the molecule. On passing electromagnetic radiation in the ultraviolet and visible region through a compound with multiple bonds, portion of the radiation is normally absorbed by the compound. The amount of absorption depends on the wavelength of the radiation and the structure of the compound.[57] Solutions of transition metal ions can be colored because d electrons within the metal atoms can be excited from one electronic state to another. The color of metal ion solutions is strongly affected by the presence of other species, such as certain anions or ligands. Molecules containing π-electrons or non-bonding electrons can absorb energy in the form of ultraviolet or visible light to excite these electrons to higher anti-bonding molecular orbitals.[58]

The Beer–Lambert law states that the absorbance of a solution is directly proportional to the concentration of the absorbing species in the solution and the path length.[59] The absorption of ultraviolet (UV) or visible light results in a change in energy of absorbing molecule. The electronic spectrum consists of bands containing several absorption lines. Each band corresponds to a definite change in the electronic energy. Individual lines within the band are due to definite vibrational transitions.

The structure of metal complexes can be predicted from the interpretation of their electronic absorption spectra and comparing them with the electronic absorption spectra of corresponding ligands. The complexes can be identified by their characteristic absorption, which is based on the position of maxima and minima in the absorption spectra along with the molar extinction coefficient value. In the spectra of transition metal ions, bands are basically of three types.

i) Bands due to d-d transitions.

ii) Charge transfer bands and

iii) Bands due to electronic transfer with the ligand.

In the transition metal complexes, visible spectra arises when an electron is excited between t_{2g} and eg orbitals with different energy levels. Charge transfer bands may arise from the transition of an electron from the ligand to the central metal atom. The molar extinction coefficient of a charge transfer band is about a hundred times greater than that of d-d transition bands.

The spectra of complexes containing higher oxidation state metal ions are more intense than the corresponding complexes of low valent metal ions. The intensity of the spectral bands sometimes reflects upon the nature of the metal-ligand bonds larger the covalency greater will be the intensity of the bands.[60]

By recording absorption spectra of a known molecule, the wavelength of radiation absorbed is correlated with the characteristic structural features. This information is then used in determining the structure of known molecules from their spectra.[61-62] The frequency at which absorption occurs depends upon the energy difference between the ground state and excited state orbital. The correct interpretation of the absorption bands in the electronic spectra gives an insight into the energy of orbitals, mode of bonding in the complexes and their geometries. By this means, it is possible to distinguish between tetrahedral and octahedral complexes and whether the shape is distorted or regular.[52]

Most organic molecules absorb in the near UV region (200-400 nm) in which the atmosphere is transparent whereas transition metal complexes absorb in the visible region (200-800 nm) because of colored solutions.

Scanning of UV / Visible Spectra of Complexes:

The synthesized Flavones and their Mn(II), Co(II), Ni(II), Cu(II), Zn(II) and Cd(II) complexes were soluble in DMSO. The UV spectra were recorded on SL159, single beam UV-VIS spectrophotometer using quartz tube of 1cm path length.

3.4 Infrared Spectroscopy:

This is one of the most widely used tools for the detection of functional groups in pure compounds and mixtures and for comparison. Infrared radiation does not have sufficient energy to cause the excitation of electrons, however, it causes atoms and groups of atoms of organic compounds to vibrate faster about the covalent bonds which connect them.[57]

There are two types of molecular vibrations stretching and bending. A stretching vibration is a rhythmical movement along the bond axis such that the interatomic distance is increasing or decreasing. A bending vibration may consist of a change in bond angle between bonds with a common atom or the movement of a group of atoms with respect to the remainder of the molecule without movement of the atoms in the group with respect to one another.[63]

In order for a vibrational mode in a sample to be "IR active", it must be associated with changes in the dipole moment. A permanent dipole is not necessary, as the rule requires only a change in dipole moment.[64]

In this spectroscopy, stretching and bending of bonds are recorded. The energy needed for stretching and bending of individual bonds, while still very small, corresponds to shorter wavelengths. These wavelengths lie in the infrared region that is, infrared radiation just to the side of visible light radiation.

When the carbon skeleton of a molecule vibrates, all the bonds stretch and relax in combination. However, some bonds stretch essentially independent of the rest of the molecule. This occurs if the bond is either.

1. Much stronger or weaker than others nearby, or
2. Between atoms that are much heavier or lighter than their neighbors.

Indeed, the relationship between the frequency of the bond vibration, the mass of the atoms and the same as Hook's law for a simple harmonic oscillator.

$$\upsilon = \frac{1}{2\pi c}\sqrt{\frac{f(m1+m2)}{m1.m2}}$$

The equation shows that the frequency of vibration υ is proportional to the square root of a force constant more or less the bond strength and inversely proportional to the root of a reduced mass μ, that is the product of the masses of the two atoms forming the bond divided by their sum.

$$\mu = \frac{m1m2}{m1+m2}$$

Stronger bond vibrates faster and so do lighter atoms.

Infrared spectra are simple absorption spectra. The sample is exposed to infrared radiation and the wavelength scanned across the spectrum. Whenever energy corresponding to specific wavelength is absorbed, the intensity of the radiation reaching a detector momentarily decreases, and this is recorded in the spectrum. Infrared spectra are usually recorded as frequency measurement called wave number (cm^{-1}) which is the inverse of the true wavelength λ in centimeters. Higher numbers are to the left of the spectrum because it is really wavelength that is being scanned.

The energy equation as-

$$E = h\upsilon = \frac{hg}{\lambda}$$

$$\text{Since } \lambda = \frac{c}{\upsilon}$$

E = Energy required to excite a bond vibration can be expressed as the inverse of a wavelength λ or as a frequency υ.

h = Planks constant.

c = Velocity of light.

The infrared frequency region above 1500 cm^{-1} in the spectrum is known as the "Functional group region". The infrared region below 1500 cm^{-1} is called "The single bond region or Fingerprint" region. Where the single bond vibration occurs that is individual bonds may vibrate independently of rest of the molecule. The bond like C-C, C-N, C-O, are characteristic in fingerprint region.[65]

IR absorption occurs not only with organic molecules but also with covalently bonded metal complexes, which are generally active in the longer wavelength of IR region. The inorganic complexes derived from the organic chelating group have a tendency to absorb in the IR region 400-4000 cm^{-1} which is of greatest practical value in the study of metal complexes. IR studies thus provide much useful information about metal complexes.

Scanning of IR Spectra:

The IR spectra of Flavones and their metal complexes of Mn(II), Co(II), Ni(II), Cu(II), Zn(II) and Cd(II) were recorded on a Perkin Elmer-Spectrum RX-IFTIR Spectrophotometer over the range 400-4000 cm^{-1} using KBr pellet technique at CIL Chandigarh, Punjab.

3.5. ^{1}H NMR Spectroscopy:

NMR requires electromagnetic waves in the radio wave region of the spectrum to make nuclei flip from one state to another. Proton nuclear magnetic resonance (^{1}H NMR) spectroscopy is concerned with the proton (^{1}H) nuclei having spin quantum number I=1/2. When sample placed in a magnetic field, NMR active nuclei such as ^{1}H or ^{13}C absorb electromagnetic radiation at a frequency characteristic of the isotope.[66] The resonant frequency, an energy of the radiation absorbed, and the intensity of the signal are proportional to the strength of the magnetic field.

^{1}H NMR allows detecting proton nuclei and saying what sort of environment is there within the molecule. It gives the relation between the number of signals in the spectrum and the number of different kinds of hydrogen atoms in the molecule. Thus one can know the different kinds of environments of the hydrogen atoms in the molecule.[57]

The atomic nuclei (^{1}H, ^{13}C) act like a tiny compass needle and have different energy levels when placed in a magnetic field. The compass needle can rotate through 360° and have essentially infinite number of different energy levels. The energy levels are quantized just like the energy levels of an electron, the energy absorption is a quantized process and the energy absorbed must equal the energy difference between the two states involved. [67]

^1H interacts and has just two different energy levels. When we apply a magnetic field to nuclei, it can either align to the lowest energy state, or it can align themselves against the field.

In the NMR spectrum, scale appears in 'parts per million' (ppm) because the exact frequency at which the nucleus resonates depends on the external magnetic field (β_0). All the protons resonate at approximately the same environment which in turn depends on its electrons and chemical environment. This approximate frequency is the operating frequency. The precise value of operating frequency is simply the frequency at which a standard reference sample resonates. A strength of magnet refers in Tesla, A 9.4 T NMR machine referred to as 400 MHz spectrophotometer.

In the positions of NMR absorption peak which arise due to the shielding or deshielding of protons by the electrons are called chemical shift. For measuring chemical shifts of various protons in a molecule, the signal for tetramethylsilane is taken as reference.[68] The reference sample generally used is tetramethylsilane, TMS ($Si(CH_3)_4$ because the four carbon atoms attached to silicon are all equivalent as silicon is more electropositive than carbon are electron rich or (shielded), which means they resonates at a frequency a little less than that of most organic compounds. This is useful because it means our reference sample is not bang in the middle of our spectrum.[65]

The chemical shift, δ, in parts per million (ppm) of a given nucleus in the sample is defined in terms of resonance frequency as:

$$\delta = \frac{\text{Frequency (Hz)} - \text{Frequency TMS ()Hz}}{\text{Operating frequency TMS (MHz)}}$$

Factors such as electronegativity, density of electron cause particular nuclei to appear at different chemical shifts (δ). The scale is from upfield for the high field (small δ=0) and downfield for a low field (large δ). The scanning of ^1H NMR spectra of ligands is done as discussed earlier in Chapter II.

3.6. Mass Spectrometry:

Mass spectroscopy is used to characterize organic molecules to measure exact molecular weights and from this exact molecular formula can be detected, and to indicate within a molecule the points at which it prefers to fragmentation.[57]

In mass spectroscopy, the molecule is bombarded with highly energetic electrons that knock a weakly bound electron out of the molecule losing a single electron leaving behind a radical cation, an unpaired electron and a positive charge. The electron that is lost will be one of relatively high energy, and this will typically not involved in bonding. If the electron beam is not too high in energy, some of these radical cations will survive the focusing operation and get to the detector. Normally two focusing operations are used. The beam is bent magnetically and electrostatically to accelerate the cations on their way to the detector and it takes about 20 μs for the cation to get there.

The strongest peak or high intense peak is assigned to an abundance of 100% and called the base peak. The intensity of other peak is shown relative to the base peak. The peak at the highest mass is the molecular ion peak corresponding to a molecular formula.

The intensity of the molecular ion peak depends on the stability of the molecular ion. The most stable molecular ions are those of purely aromatic systems. If substituents that have a favorable mode of cleavage are present, the molecular ion peak will be less intense and the fragment peak relatively more intense.[63]

The fragments produced may be useful in identifying the molecule. Mass spectroscopy separates isotopes. The isotopes occur one or two mass unit above the mass of the normal atom. Therefore, besides looking for the molecular ion peak, to locate M^{+1} and M^{+2} peaks.[67]

In the present study, the mass spectra off favones and their metal complexes are presented in Chapter II.

3.7. Thermal Analysis:

The technique in which change in physical and chemical properties of a substance is measured as a function of temperature when subjected to a controlled temperature program is referred to as thermoanalytical techniques.

These thermal methods are based upon the measurement of the dynamic relationship between temperature and some property of a system such as mass, the heat of reaction or volume.[69]

Thermogravimetric (TG) Analysis:

TG provides the analyst a quantitative measurement of any weight change associated with a transition. Thermogravimetry is a technique in which the change in the weight of a substance when heated or cooled at a controlled rate is recorded as a function of temperature or time. In TG generally, the mass of a sample is recorded continuously as its temperature is increased linearly from room temperature to as high as 1000 or 1200°C with an inert or reactive atmosphere. A plot of sample weight change as a function of temperature referred to as "Thermogram" provides both qualitative and quantitative information of the sample. It can directly record the loss in weight as a function of temperature for transitions that involve dehydration or decomposition. Oxidative mass losses are the most common observable losses in TGA.[70]

Thermogravimertic curves are characteristic of a given compound or material due to the unique sequence of physical changes that occur over definite temperature ranges. Changes in weights are the results of breaking or formation of various bonds and evaluation of volatile or heavier reaction products. The record of the TG curve the mass should be plotted on the ordinate decreasing downwards and temp (T) or time (t) on the abscissa increasing from left to right. The heating rate is the rate of temp increase which is customarily quoted in degree per minute (on the Celsius).

Kinetic study of thermal decomposition is important in calculating various parameters like the energy of activation (Ea), free energy change (ΔG), pre-exponential factor (Z), order of reaction (n), and entropy of activation (ΔS), which directly govern the factors of thermal stability of any substance. The apparatus required for a thermogravimetric analysis includes.

i) A sensitive recording thermobalance.

ii) Furnace

iii) Furnace temperature controller and programmer

iv) Recorder to provide a plot of a sample mass versus temperature and auxiliary equipment to maintain an inert atmosphere for the sample. The course and character of the TG curves are influenced by several experimental factors such as the rate of heating, the atmosphere of the furnace, the geometry of the furnace, sample holder, particle size, and heat of reaction and compactness of sample.[71]

In TG operation the resulting electrical current is proportional to the weight change and is recorded. Change in weight can also be detected by contraction or elongation of a precision vehicle spring its movement detected by electrically. The furnace temp is continuously monitored and linear heating rates from 5 to 20^0C/min are generally employed.

Differential Thermal Analysis:

Differential Thermal Analysis (DTA) is a technique in which the difference in temperature between a sample substance and an thermally inert reference material (α- alumina, silicon carbide, glass beads) is recorded as a function of sample temperature or time as the two specimens are subjected to identical temperature regimes in an environment heated or cooled at a controlled rate which is linear with time.[72]

Differential thermal analysis is a thermoanalytical technique that is similar to differential scanning calorimetry. In DTA, the material under study and an inert reference are made to undergo identical thermal cycles, while recording temperature difference between sample and reference.[73] DTA component shows whether decomposition processes are endothermic or exothermic. The DTA also measures temp corresponding to phase change where no mass loss occurs such as melting, recrystilization and glass transitions.

Any transition which the sample undergoes will result in liberation or absorption of energy by the sample with a corresponding deviation of its temperature from that of reference. The differential temperature (DT) versus the programmed temperature of transition, and whether the transition is exothermic or endothermic is shown. Each substance will give a DTA curve whose number, size, shape and position of various endothermic and exothermic features serve as a means of qualitative identification of the substance and give important information about the nature of test sample.

The area under a DTA peak can be due to the enthalpy change and is not affected by the heat capacity of the molecule. In DTA, the peaks appearing below zero are endothermic changes and those appearing above zero are for exothermic changes. Generally, phase transitions, dehydration reactions and some decomposition reactions produce endothermic effects. The heat effects occurring during these chemical and physical changes are measured by differential thermal analysis.

The endotherms and exotherms can be used to calculate the heat of reaction or heat of phase transition. The area (A) of DTA peak depends on the sample mass (m), enthalpy change (ΔH) of the process, calibration factors involving sample geometry (G), constant (k) related to the thermal conductivity of the sample given by

$$A = \frac{\pm Gm\Delta H}{k}$$

Where,

$$\Delta H = \text{-ve} \rightarrow \text{exothermic}$$
$$= \text{+ve} \rightarrow \text{endothermic.}$$

DTA is used to study the thermal stability of a large number of metal complexes and also finds extensive applications in determining the purity, metal-ligand bond strength and melting points.[74-69] The simultaneous use of TGA and DTA for the study of metal complexes provides many benefits. The TG, a valuable tool in its own right is perhaps most useful when it complements DTA studies. The similarity in the values of thermodynamic and kinetic parameters indicates that the basic steps involved in the thermal degradation of chelates are the same.[72] The simultaneous use of TGA and DTA plays an important role in the study of thermal behavior of inorganic substances.

Scanning of TG-DTA Curves:

The simultaneous thermogravimetric, differential thermal analysis of representative metal complexes was performed in an inert nitrogen atmosphere on Perkin Elmer STA 6000 at STIC, Cochin, Kerala. The heating rate was 10°/min and the flow rate of nitrogen 50 ml/min. The reference substance used was α Al_2O_3 in a platinum crucible and sample weighed in the range of 4-12 mg.

3.8. X-Ray Diffraction:

About 95% of all solid materials can be described as crystalline. When X-ray interacts with crystalline substance one gets a diffraction pattern. X-ray diffraction is a powerful method to understand the structure of the compound.[75-77] The structure and symmetry properties of complexes can be determined by this method. It gives information of inter atomic distance. The X-ray diffraction pattern of a pure substance is therefore like a fingerprint of the substance. When the beam of monochromatic X-ray strike the plane of atoms in the crystal, diffraction phenomenon is observed. Relative to other methods of analysis, powder diffraction allows for rapid, non-destructive analysis of multi-component mixtures without the need for extensive sample preparation.[78]

Bragg's equation gives the relation between interplanar distance, diffraction angle and wavelength.

Where,

n = Order of reflection

λ = Wavelength of X-rays

d = Interplanar distance and

θ = Incident angle.

Bragg's law state that the constructive interference of the radiation from successive planes occurs when the path difference is an integral multiple numbers (n) of wavelength (λ). All crystals of a substance possess the same elements of symmetry.

Bragg pictured the diffraction effect or a reflection of X-rays from lattice planes of the crystals. The atoms in the plane are responsible for the scattering of X-rays.

Deby-Scherrer and Hull discovered the powder XRD technique for analyzing the powdered polycrystalline sample. The crystalline substances are powdered such that the enormous number of small crystallites is oriented in every possible direction, When an X-ray beam travels the material, a significant number of particles are expected to be oriented in such way to fulfil Bragg's condition for refection from every possible interplanar spacing. All possible lattice planes are present in a powder sample and when a beam of characteristic radiation is directed at the sample, the Bragg's condition for reflection is simultaneously satisfied for all types of lattice planes. In powder method, a continuous cone of diffracted rays is produced and these cones are determined by spacing of prominent planes. The characteristic features of X-ray powder diffractogram are the angles at which reflections occur and the relative intensities of the reflections. The refection angle θ depends on the wavelength λ of X-ray and the inter planar spacing d_{hkl} in accordance with Bragg's law. The spacing d depends on the dimensions of the unit cell. The relative intensities depend on the number, kind and arrangements of atoms in the lattice planes.

The powder diffractogram of an unknown compound consists, of measuring the diffraction angles θ or 2θ. Once θ is determined one can calculate the interplanar spacing d of reflecting planes from the values of λ of X-rays used and then the dimensions of the unit cell are deduced. We can easily determine the unit cell of compounds with lower symmetry like orthorhombic, monoclinic, and triclinic which are characterized by 3, 4 or 6 lattice constant respectively.

In spite of these limitations, the powder XRD is an important technique for the identification of unknown compounds because of applicability to all crystalline substances and its very convenient way of preparation of the sample and its presentation to the beam. The powder diffraction method is ideally suited for characterization and identification of polycrystalline phases. Hence the powder XRD method is used in the present investigation for crystal structure determination.

Scanning of X-ray Diffractogram and Indexing of the Powder Data:

The X-ray diffractograms of representative metal complexes were scanned on Bruker AXS D_8 Advance attached to a digital computer along with graphical assembly in which Cu-K_α radiation source connected with tube Cu-Ni. The X-rays produced at 40kV generator tension, Generator current ratio of 0.5 were used to scan the diffractogram over an apex angle 2θ range from 0-60° with scan speed, 0.055° (2θ) per second employing proportional counter. Every reflection position and reflection height was recorded in terms of relative intensity. The area under the peak is proportional to the intensity of the reflection. The position of a diffraction peak is 'independent' of the atomic positions within the cell and entirely determined by the size and shape of the unit cell of the crystalline phase. Each peak represents a certain lattice plane and can therefore, be characterized by a Miller index.

The obtained data was indexed using computer program PDP-Burn. The preliminary data in the form of 2θ and intensity ratio were fed to the computer and corresponding h, k, l values were assigned to each peak. The program calculates the lattice parameters a, b, c (A°) and α, β, γ (degree) along with standard deviation in each. The crystal volume is also obtained from the refined cell data.

The pyknometric densities of each of the complexes were determined using specific gravity bottle by the principle of Archimedes.[71] The density is related to the formula factor as follows

$$D = \frac{ZFM}{V}$$

Where,

Z = Number of formula unit in the cell.

F = Formula weight of the compound.

M = Mass of a hydrogen atom in amu.

V = Crystal volume.

The calculated densities were correlated with experimental densities. The observed densities were used to fix the value of Z to the nearest whole number. By using literature values for definite crystal system, the probable space group was assigned to each of the complexes.[79] The Porosity of complexes also calculated by taking the difference between observed density and calculated densities.

CHAPTER 4

Infrared Spectroscopy

4.1. Introduction:

Infrared spectroscopy is a very important technique for the identification of functional groups and bonding sequences in the compounds by the absorption of light in the infrared region of the electromagnetic spectrum. In this spectroscopy, stretching and bending of bonds are recorded. For stretching and bending of bonds in the molecule very small amount of energy required, corresponding to larger wavelengths. These wavelengths lie in the infrared region which lies between the microwave region and visible region of the electromagnetic spectrum. Infrared radiations, like any electromagnetic radiations, are characterized by properties of frequency (υ) and wavelength (λ) that are related by the speed of light (c).

$$\upsilon = \frac{c}{\lambda}$$

Infrared spectra are usually recorded as frequency measurement called wave number (cm^{-1}) which is inverse of the wavelength λ. Generally, infrared spectra recorded in the region 4000-400 cm^{-1}. The infrared absorption occurs not only in organic molecule but also with covalently bonded metal complexes. Infrared spectra are used to study the surrounding environment of organic ligands and their bonding with the metal ion. The infrared spectra of free ligands are compared with that of infrared spectra of metal complexes. The infrared spectra of metal complexes differ by the corresponding stretching frequencies of original bands in the ligands, generally stretching frequencies decreases.

Formation of metal complexes can be confirmed by taking into account the following consideration. i) Change in position of bands. ii) The disappearance of old bands and the appearance of new bands. iii) Change in the intensities of bands.

The change in the position of a band is observed due to change in stretching vibration mode of bond involving coordinated atom. Additional bonds on chelation favor the appearance of new peaks. Replacement of a bond by newer one causes replacement of earlier peaks by new peaks.

The infrared spectra of flavone and metal complexes are studied in the present investigation. The different types of bonding in metal complexes were also investigated from the vibrational spectra. Chemical bonds in different environments will absorb varying intensities and varying frequencies. The neighbouring bonded groups also influenced this vibrational frequency.

The infrared spectra of flavones and their metal complexes were recorded on a Perkin-Elmer Spectrum RX-IFTIR Spectrophotometer over the range 4000-400 cm^{-1} using KBr pellet at CIL, Chandigarh, Punjab.

4.2. Infrared Spectral Studies of Flavones:

Infrared Spectral Studies of Flavones:

Enolic (-OH) Stretching Frequency:

Elfi Susanti VH et al.[80] have assigned a broad weak band at 3464 cm^{-1} in the flavone which is synthesized from 2,6-dihydroxy acetophenone and 3,4 di-methoxybenzaldehyde. Omar H. Al-Obaidi et al.[81] was assigned υ (-OH) stretching vibration band at 3426 cm^{-1} for Ar-OH of flavones. Mokle S. S. et al.[82] have assigned a broad weak band at 3405 cm^{-1} for υ (-OH) stretching vibration band in the flavones. Jingliang Zhang et al. were reported υ (-OH) stretching vibration band at 3423 cm^{-1}.

The flavones in the present investigation has one (-OH) group at first position of flavone A ring exhibit υ (-OH) stretching vibration band in the range of 3405-3464 cm^{-1} and are agreement with the reported data.

Carbonyl group (-C=O in Pyron ring) Stretching Frequency:

R. A. Shastri[83] shows that the structure of flavones gives IR band at 1595 cm^{-1} for (-C=O in pyron ring) and the absence of –OH group confirmed the oxidation of chalcone into the flavones. Omar H. Al-Obaidi et al.[81] was assigned υ (-C=O) stretching vibration band at 1641 cm^{-1}. Venkatachalam et al.[84] assigned the flavones to show strong absorption band at 1606-1616 cm^{-1}corresponding to the υ (-C=O) stretching frequency. Mokle S. S. et al.[82] have assigned a broad weak band 1639 cm^{-1} for υ (-C=O) stretching frequency in the flavones.

(C-O-C) Stretching Frequency:

R. A. Shastri[83] shows that the flavones give IR band at 1230$cm^{-1}$for (C-O-C) Stretching Frequency. This was synthesizing from O-hydroxyacetophenone and furan-2-Carbaldehyde. Elfi Susanti VH[1] et al.[80] have assigned a broad weak band at 1273 $cm^{-1}$for (C-O-C) Stretching Frequency in the flavone which is synthesized from 2,6-dihydroxy acetophenone and 3,4 di-methoxybenzaldehyde. Y.Murti et al.[85] assigned (C-O-C) Stretching Frequency at 1178 cm^{-1} in the flavonones.

(C=C) Stretching Frequency:

R.A. Shastri[83] shows that the flavones gives IR band at 1658 $cm^{-1}$for (C=C) Stretching Frequency. This was synthesizing from O-hydroxy acetophenone and furan-2-Carbaldehyde. Y. Murti et al.[85] assigned (C=C) Stretching Frequency at 1598 cm^{-1} in the flavonones. M.M. Rathore et al.[86] reported (C=C) Stretching Frequency at 1609 cm^{-1}. Heena Kausar et al.[87] assigned (C=C) Stretching Frequency at 1565-1595 cm^{-1} lowering of carbonyl frequency is attributed to the conjugation with carbonyl group.

Elfi Susanti VH et al.[80] have assigned a broad weak band at 1589-1465 cm^{-1} for aromatic (C=C) Stretching Frequency in the flavone which is synthesized from 2,6-dihydroxy acetophenone and 3,4 di-methoxybenzaldehyde. Mokle S. S. et al.[82]

have assigned a broad weak band 1570-1489 cm^{-1} for υ (-C=C) stretching frequency in the aromatic rings of flavones. Y.Murti et al.[85] assigned (C=C) Stretching Frequency at 1567, 1485, 1443 cm^{-1} in aromatic rings of the flavonones.

Other Stretching Frequencies:

R.A. Shastri[83] shows that the flavones give IR band at 3110 cm^{-1}symmetric and 3065 cm^{-1}asymmetric for Ar-H Stretching Frequency. This was synthesizing from O-hydroxyacetophenone and furan-2-Carbaldehyde. Y. Murti et al.[85] assigned (Ar-H) Stretching Frequency at 3012 cm^{-1} in aromatic rings of the flavonones.

Omar H. Al-Obaidi et al.[81] was assigned υ (Ar-H) stretching vibration band at 3075 cm^{-1}. M.M. Rathore et al.[86] reported (C-H) Stretching Frequency for alkyl group at 2864-2979 cm^{-1}.

Table 4.1 Salient Features of IR Spectral Data of Flavones. (Assignment of band frequencies to bond vibration modes)

Ligands	υ(OH) Enolic Stretch freq in cm^{-1}	(-CO-CH=CH-) α, βunsaturated carbonyl Stretch freq in cm^{-1}	Carbonyl group (-C=O) in pyron ring) Stretch freq in cm^{-1}	(C-O-C) Stretching frequency in cm^{-1}	(C=C) Stretching frequency in cm^{-1}	Aromatic Ring (C=C) Stretching frequency in cm^{-1}	Ar-H Stretching frequency in cm^{-1}	$-NO_2$ Stretching frequency in cm^{-1}
B_1	3516	-	1605	1232	1502	1456	3121	-
B_2	3011	-	1606	1204	1550	1430	2919	-

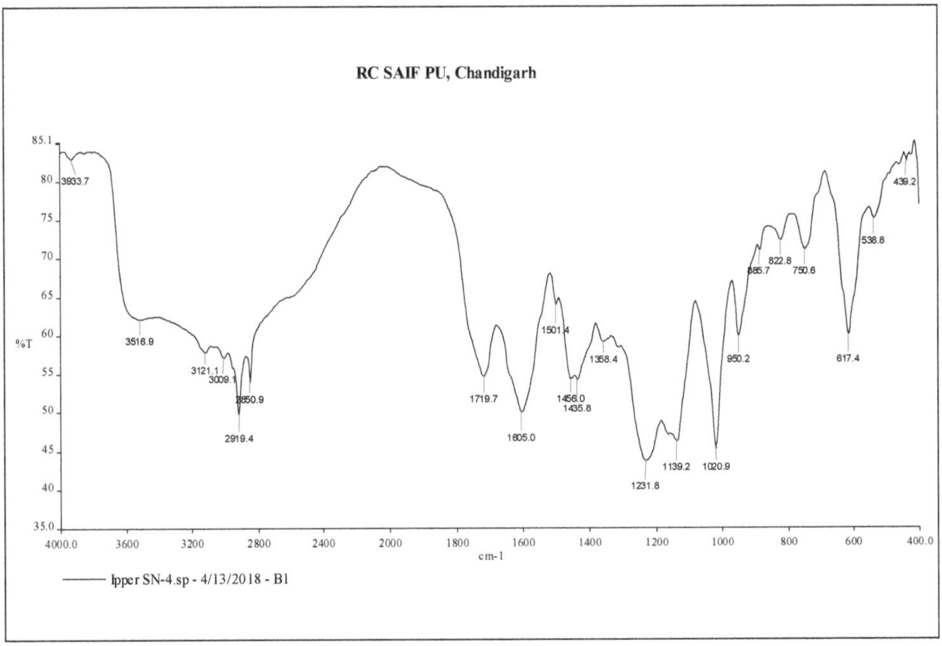

Fig. 4.1: Infrared Spectrum of Flavone B1

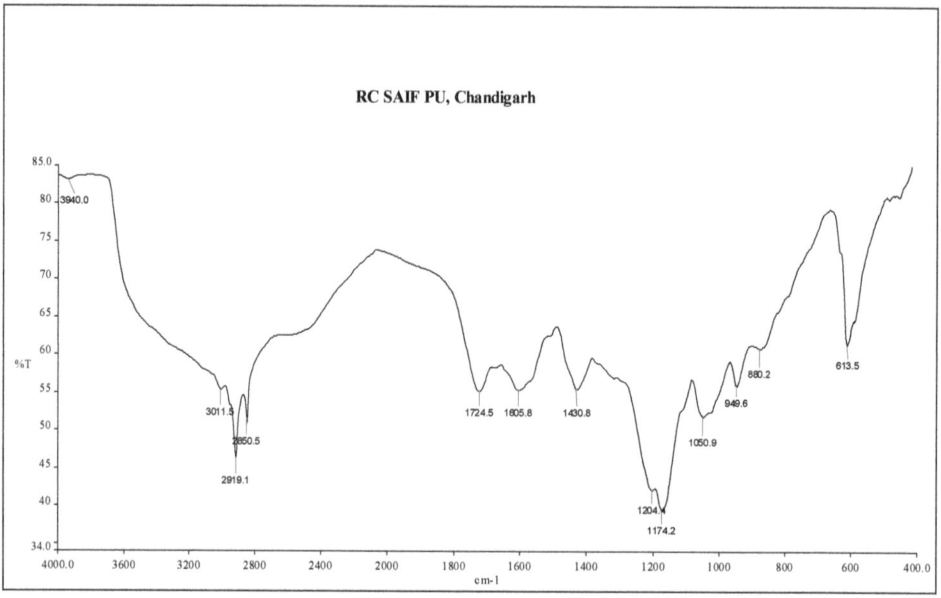

4.2 Infrared spectrum of Flavone B2

4.3. Infrared Spectral Studies of Metal Complexes:

In B_1 flavone ligand, there is a presence of phenolic –OH is confirmed by peak at 3516 cm^{-1} in the spectra of Mn(II) complexes, that is the complete disappearance of the peak at 3516 cm^{-1} suggests the absence of phenolic group –OH indicates its coordination. (C-O-C) is shifted to a lower wave number compared with a free ligand.

The new band is observed at 3363 cm^{-1} which is due to the -OH intense broad band of coordinated water molecules. In Mn(II) complexes new band is observed at 661 cm^{-1} due to the (M-O) bond.

In the spectra of Mn(II) complexes, the strong bands appeared in the range of 1502 cm^{-1} can be assigned to (C=C) stretching vibrational mode. The bands due to (C=C) aromatic ring vibrations may remain the same.

Table 4.2 : Infrared Absorption Frequencies (cm^{-1}) of Mn(II) Complexes.

(Assignment of band frequencies to bond vibration modes)

Ligand/ Metal complexes	υ (OH) cm^{-1}	υ (H_2O) cm^{-1}	υ (-CO-CH=CH-) cm^{-1}	υ (-C=O in pyron ring) cm^{-1}	υ (C-O-C) cm^{-1}	υ (C=C) cm^{-1}	Aromatic Ring (C=C) cm^{-1}	υ (M-O) cm^{-1}
B_1	3516	-	-	1605	1232	1502	1456	-
$[Mn(B_1)_2]$	-	3363	-	1574	1019	1431	1234	661

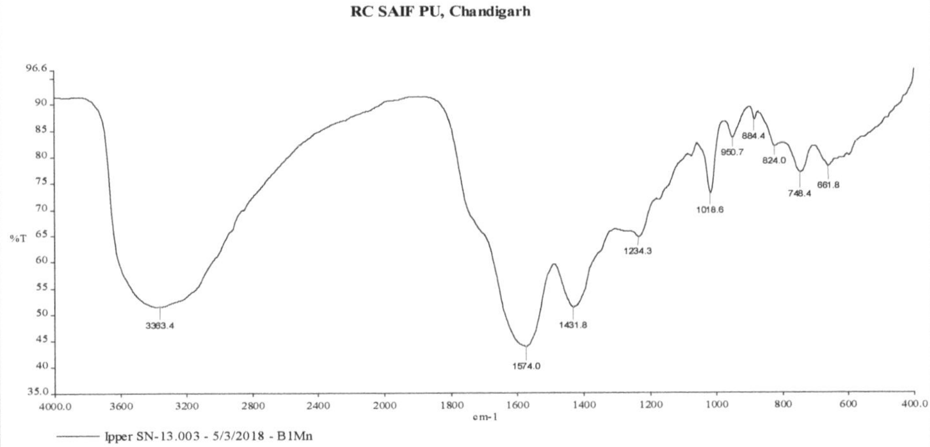

Fig. 4.3: Infrared Spectra of Mn(II) Complex of Flavone B_1.

Infrared Spectral Studies of Ni(II) Complexes:

In B_2 flavone ligand, there is the presence of phenolic –OH which is confirmed by peaks at 3011 cm^{-1} in the spectra of Ni(II)) complexes, that is the complete disappearance of peak at 3011 cm^{-1} suggests the absence of phenolic group –OH indicates its coordination. (C-O-C) is shifted to a lower wave number comparing with free a ligand.

The new band is observed at 3169 cm^{-1} which is due to the -OH intense broad band of coordinated water molecules. In Ni(II)) complexes new band is observed at 483 cm^{-1} due to the (M-O) bond.

In the spectra of Ni(II)) complexes, the strong bands appeared in the range of 1397 cm^{-1} can be assigned to (C=C) stretching vibrational mode. The bands due to (C=C) aromatic ring vibrations may remain the same.

Table 4.3 : Infrared Absorption Frequencies (cm^{-1}) of Ni(II) Complexes.

(Assignment of band frequencies to bond vibration modes)

Ligand/ Metal complexes	υ (OH) cm^{-1}	υ (H_2O) cm^{-1}	υ (-CO-CH =CH-) cm^{-1}	υ (-C=O in pyron ring) cm^{-1}	υ (C -O -C) cm^{-1}	υ (C = C) cm^{-1}	Aromatic Ring (C=C) cm^{-1}	υ (M-O) cm^{-1}
B_2	3011	-	-	1606	1204	1550	1430	-
$[Ni(B_2)_2]$	-	3169		1620	1199	1397	-	483

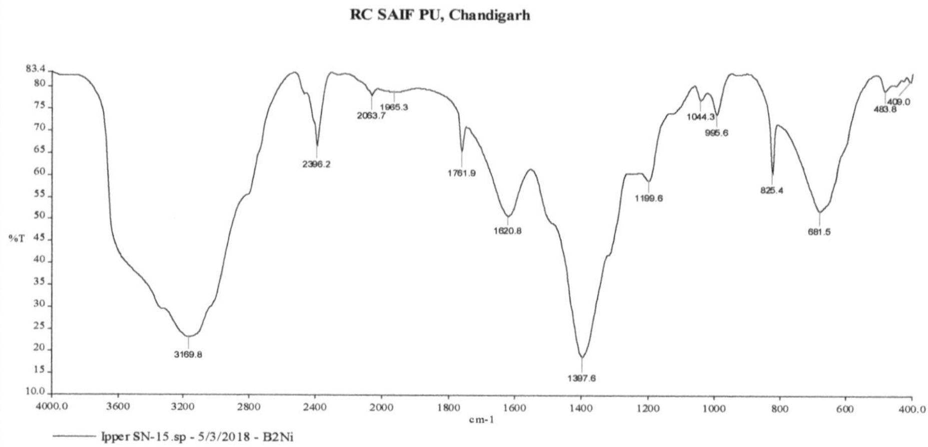

Fig. 4.4: Infrared Spectra of Ni(II) Complex of Flavone $B_{2.}$

Infrared Spectral Studies of Cu(II) Complexes:

In B_1 flavone ligand, there is the presence of phenolic -OH is confirmed by peaks at 3516 cm^{-1} in the spectra of Cu(II)) complexes, that is the complete disappearance of the peak at 3516 cm^{-1} suggests absence of phenolic group -OH indicates its coordination. (C-O-C) is shifted to a lower wave number comparing with a free ligand.

The new band is observed at 3189 cm^{-1} which is due to the -OH intense broad band of coordinated water molecules. In Cu(II)) complexes new band is observed at 456 cm^{-1} due to the (M-O) bond.

In the spectra of Cu(II)) complexes, the strong bands appeared in the range of 1434 cm^{-1} can be assigned to (C=C) stretching vibrational mode. The bands due to (C=C) aromatic ring vibrations may remain the same.

Table 4.4: Infrared Absorption Frequencies (cm^{-1}) of Cu(II) Complexes.

(Assignment of band frequencies to bond vibration modes)

Ligand/ Metal complexes	**υ (OH) cm^{-1}**	**υ (H_2O) cm^{-1}**	**υ (-CO-CH=CH-) cm^{-1}**	**υ (-C=O in pyron ring) cm^{-1}**	**υ (C-O-C) cm^{-1}**	**υ (C=C) cm^{-1}**	**Aromatic Ring (C=C) cm^{-1}**	**υ (M-O) cm^{-1}**
B_1	3516	-	-	1605	1232	1502	1456	
$[Cu(B_1)_2]$	-	3189		1603	1142	1434		456

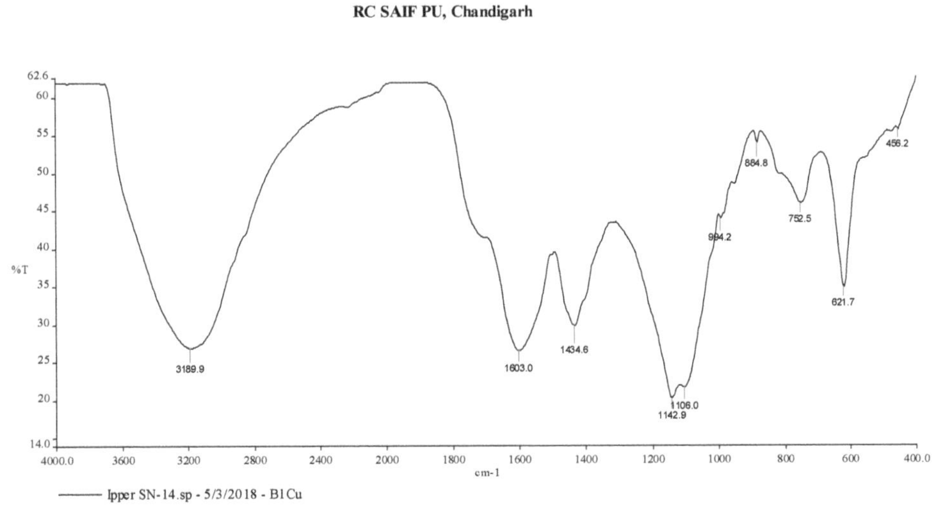

Fig. 4.5: Infrared Spectra of Cu(II) Complex of Flavone B1

4.4. Conclusions:

From the above discussion on IR spectra data of flavones and their metal chelates, the following conclusions may be drawn.

The flavones in the present investigation have one (-OH) group at first position of flavone A ring exhibit υ (-OH) stretching vibration band in the range of 3405-3464 cm^{-1} and are in agreement with the reported data. The structure of flavones gives IR band at 1595 cm^{-1} for (-C=O in pyron ring) and the absence of –OH group confirmed the oxidation of chalcone into the flavones. The flavones give IR band at 1230 $cm^{-1}$for (C-O-C) Stretching Frequency. The flavones gives IR band at 1658 cm^{-1} for (C=C) Stretching Frequency.

In the flavone ligand, there is a presence of phenolic –OH which is confirmed by peak at 3405-3464 cm^{-1} in the spectra of metal complexes, that is the complete disappearance of the peak. This suggests the absence of phenolic group -OH indicates its coordination. (C-O-C) is shifted to a lower wave number comparing with a free ligand.

The new band is observed at 3189 and 3363 cm^{-1} which is due to the -OH intense broad band of coordinated water molecules. In metal complexes, the new band is observed at 661 cm^{-1} due to the (M-O) bond.

In the spectra of metal complexes, the strong bands appeared in the range of 1502 cm^{-1} can be assigned to (C=C) stretching vibrational mode. The bands due to (C=C) aromatic ring vibrations may remain the same.

On the basis of results of elemental analysis, NMR and Mass spectral data it may be concluded that the complexes are having octahedral structures except for Zinc which has a square planar structure with 1:2 metal to ligand ratio.

The following probable structures may be proposed to the metal chelates of bidentate flavone ligands in the present study are represented in fig.

Metal complexes of flavones B1, B2 R=H, -CH3
M= Mn(II), Co(II), Ni(II), Cu(II) and Cd(II)

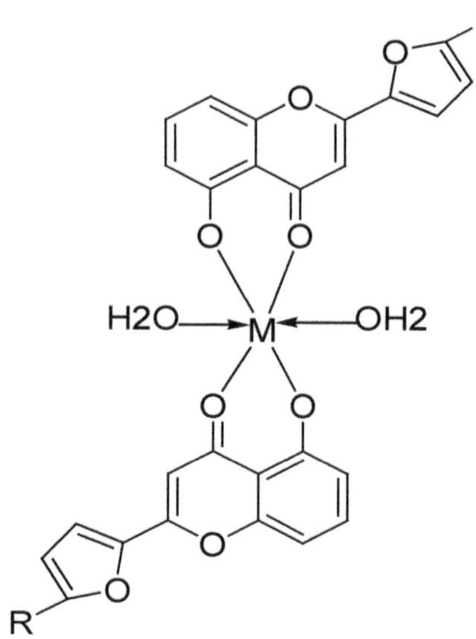

Zinc Flavone Metal Complexes

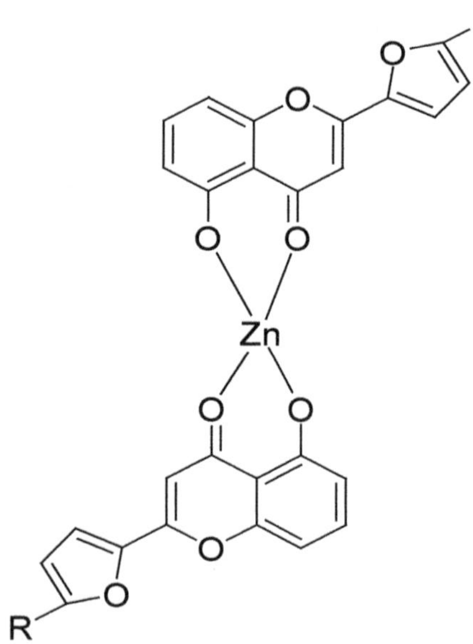

CHAPTER 5

Elemental Analysis, Solution Conductivity, Magnetic Susceptibility and Electronic Absorption Spectral Studies of Metal Complexes

5.1. Manganese (II) Complexes:

Theoretical Considerations:

The Mn(II) complexes have d^5 configuration. The divalent state of manganese is most stable and most the important oxidation state. The most common geometry of Mn(II) complex is octahedral geometry. The Mn(II) complexes also exist in other geometrical forms such as a tetrahedral, square planar, 5-coordinate and 7-coordinate. The octahedral Mn(II) complexes are more stable than the other geometrical forms.

Mn(II) complexes have high spin behavior and hence show low ligand field stabilization energy. The high spin Mn(II) complexes are expected to have the spin only magnetic moment of 5.92 B.M. because the crystal field or spin orbit coupling alone is unable to list the sixfold degeneracy of the $^6A_{1g}$ state and five unpaired electrons. The magnetic moment of octahedral and tetrahedral high-spin complexes are close to each other i.e. 5.92 B.M. The low spin octahedral complexes are having $^2T_{2g}$ ground term having magnetic moment 2.5 B.M.[88]

Many high spin octahedral Mn(II) complexes are very light colored and intensities of transition are extremely low than the bands of normal d-d transitions. All excited states of the d^5 system have different spin multiplicity from the ground state, the transitions to these states are spin orbit interactions, they are not totally absent but give very weak absorption bands.

The four lowest energy bands in the spectra of high spin octahedral Mn(II) complexes may be assigned to the transition give below.[89]

17,000-18,000cm^{-1}	$^6A_{1g} \rightarrow {}^4T_{1g}(G)$
21,000-23,000cm^{-1}	$^6A_{1g} \rightarrow {}^4T_{2g}(G)$
25,000-27,000cm^{-1}	$^6A_{1g} \rightarrow {}^4E_{1g}(G)$
	$^6A_{1g} \rightarrow {}^4A_{1g}(G)$

Elemental Analysis and Solution Conductivity:

In the present investigation, all the Mn(II) complexes synthesized were dark brown in color and stable to air and moisture at room temperature. The complexes decompose at high temperature without melting. They are insoluble in water but soluble in DMSO and DMF.

The observed molar conductivities of Mn(II) complexes are reported in Table 5.1, The low solution conductivity of 10^{-3} M solutions of Mn(II) complexes in DMSO indicates their non-electrolytic nature.

The elemental analysis data (Table 2.13) and physical parameters of Mn(II) complexes are given in Table 2.7, reveals that the Mn(II) complexes have 1:2 metal to ligand ratio in monomeric form.

Magnetic Susceptibility and Electronic Absorption Spectral Studies:

The magnetic moment of Mn(II) complexes in the present investigation are in the range which is almost close to the spin only value of 5.2 B.M. These values are in good agreement with the moment reported for mononuclear high spin octahedral Mn(II) complexes by earlier workers.[90]

The observed magnetic moments of Mn(II) complexes are given in Table 5.1.

Table 5.1 : Solution Conductivity, Magnetic Susceptibility and Electronic Absorption Spectral Data of Mn(II) Complexes.

Mn(II) Complexes	Molar Conductance Ohm^{-1} cm^2mol^{-1}	μ_{eff} (B.M.)	Absorption Maxima cm^{-1} (nm)		
			$^6A_{1g} \rightarrow ^4T_{2g}$ (G)	$^6A_{1g} \rightarrow ^4A_{1g}$ (G), ^4E_g	Charge Transfer
B_1	6.97	5.97	24154(414)	27624(362)	29673(337)
B_2	6.70	5.48	24570(407)	27777(360)	29154(343)

Omar H. Al. Obaidi studied Mn(II) complexes of flavones the electronic absorption at 33333, 20408, and 37735 cm^{-1} and magnetic moment 5.84 B.M. suggested octahedral environment around Mn(II) ion. The transitions assigned to the above bands were $^6A_{1g} \rightarrow ^4T_{2g}$(G), $^6A_{1g} \rightarrow ^4A_{1g}$ (G) and $^6A_{1g} \rightarrow ^4E_{1g}$(G) respectively.[91]

Faize-Beg and et al.[92] shows that electronic spectra of Mn(II) complexes of flavone chrysin gives charge transfer band at 24691 cm^{-1} (405 nm).

The electronic absorption spectra of Mn(II) complexes in DMSO solution show very weak band which cannot be detected clearly. The magnetic moment values for Mn(II) having 5.5-5.9 B.M. These values consist of high spin octahedral complex.[93-94]

Based on the inferences drawn from the literature on analogous compounds and the interpretations on the conductivity, element and metal analysis, magnetic moment and electronic spectral data of complexes, a mononuclear structure with octahedral geometry may be suggested to Mn(II) complexes in the present work.

5.2. Cobalt (II) Complexes:

Theoretical Considerations:

Werner and Jorgensen say that cobalt complexes have interesting structural features for the development of coordination chemistry. The most common oxidation states for cobalt are +2 and +3, where the +2 and +3 ions resulted in the formation of paramagnetic and diamagnetic complexes.

The high spin Co(II) ion have d^7 configuration. The high spin octahedral complexes show the magnetic moment up to 5.2 B.M. at room temperature. The spin only value being 3.8 BM for three unpaired electrons for low spin octahedral Co(II) complexes possessing ^2E_g ground state term, the observed magnetic moments are in the range of 1.8-1.9 B.M. The high spin octahedral Co(II) complexes possess $^4T_{1g}$ ground state term.

Elemental Analysis and Solution Conductivity:

In the present investigation, all the Co(II) complexes are brown in color, stable to air and moisture. Decomposes at high temperature rather than showing sharp melting points. They are insoluble in water and soluble in DMSO, DMF. The low conductivity values in DMSO solution (10^{-3} M) are given in Table 5,2 indicates non-electrolytic nature. Elemental analysis data reveals that the observed percentages of C,H,N and metal ion are in good agreement with values predicted and calculated for Co(II) complexes with 1:2 metal to ligand ratio (Table-2.8 & 2.14).

Magnetic Susceptibility and Electronic Absorption Spectra of Co (II) Complexes

In the present investigation, all the Co(II) complexes contain a paramagnetic d^7 species. All complexes showed magnetic susceptibility values. This observation is in agreement with previously reported magnetic properties.[95] The observed magnetic moments of the complexes are presented in Table 5.2.

Table 5.2 : Solution Conductivity, Magnetic Susceptibility and Electronic Absorption Spectral Data of Co(II) Complexes.

Co(II) Complexes	Molar Conductance $Ohm^{-1}cm^2mol^{-1}$	μ_{eff} (B.M.)	Absorption Maxima cm^{-1} (nm)	
			$^4T_{1g(F)} \rightarrow {}^4T_{1g(P)}$	Charge Transfer
B_1	6.76	4.19	32258 (310)	27855 (359.5)
B_2	7.89	4.23	29069 (344)	27100 (369)

Omar H.Al.Obaidi studied Co(II) complexes of flavones the electronic absorption at 38610, 33670, and 26525 cm^{-1} suggested octahedral environment around Co(II) ion. The transitions assigned to the above bands were $^4T_1 \rightarrow {}^4A$, $^4T_1 \rightarrow {}^4T_{(P)}$, $^4T_1 \rightarrow {}^4T_2$ and $^4T_1 \rightarrow {}^2E$ respectively. The observed magnetic moments of 4.17 B.M. suggested octahedral environment around Co(II) ion.[91]

On the basis of inferences drawn from literature and interpretation of data, element analysis, magnetic and electronic absorption spectral study of complexes, octahedral geometry may be proposed to Co(II) complexes in the present study.

5.3. Nickel (II) Complexes:

Theoretical Considerations:

The Ni(II) has $3d^8$ configuration. The main structural types of Ni(II) are octahedral, tetrahedral and square planar. Strong ligand fields pair up the unpaired of Ni(II) complexes giving rise to the formation of diamagnetic square planar or trigonal bipyramidal complexes. Weak ligands give rise to outer octahedral complexes, paramagnetic in nature, which are generally blue or green colored. Square planar complexes are red, yellow or brown in color. The magnetic moment values of octahedral Ni(II) complexes fall in the range 2.9 to 3.4 B.M. These values of magnetic moment are slightly higher than expected from spin only values of 2.83 B.M. due to spin-orbit coupling and contribution from the ^3A_2g And $^3T_{2g}$ state. The magnetic moment of tetrahedral geometries are in the range of 3.5 to 4.0 B.M. indicating higher orbital contribution due to an orbitally degenerate $^3T_1(P)$ term of tetrahedral ligand fields of Ni(II) complexes.

Elemental Analysis and Solution Conductivity:

The newly synthesized Ni(II) complexes in the present investigation are green in color. They are stable to air and moisture. The complexes decompose at a high temperature. It is observed that the Ni(II) complexes are insoluble in water but soluble in DMF and DMSO. The solution conductivity values of these complexes in 10^{-3} M DMSO show that they are non-electrolyte in nature. The observed conductivity values of Ni(II) complex are presented in Table 5.3. The elemental analysis data (Table 2.15 & 2.9) indicate that the observed percentage of C, H, N and metal ion are in good agreement with the values predicted and calculated assuming 1:2 metal to a ligand ratio corresponding to monomeric nature of Ni(II) complexes.

Magnetic Susceptibility and Electronic Absorption Spectra of Ni(II) Complexes:

All the Ni(II) complexes in present investigation are paramagnetic in nature. This indicates the octahedral geometry of the complexes. In the present investigation electronic spectra of the Ni(II) complexes show d-d band in the range 23640-27027 cm^{-1} assigned to the paramagnetic $^3A_{2g} \rightarrow {}^3T_{1g}$ (P) spin allowed transition, consistent with their octahedral configuration. The electronic spectra of Ni(II) complexes also show charge transfer band in the range of 29761 to 37735 cm^{-1}. Khushal G. Karamunge and et al.[96] have synthesized complexes of Ni(II) with bands at 13315-13550 and 24270-25060 cm^{-1} due to $^3A_{2g}$ (F)$\rightarrow {}^3T_{1g}$ (F) and $^3A_{2g}$ (F)$\rightarrow {}^3T_{1g}$ (P) transition respectively and having magnetic moment 3.1 to 3.13 B.M. range. This indicates that the complex of Ni(II) has octahedral geometry. Reddy Vidyavati et al.[53] reported that the Ni(II) complexes exhibit three bands in the region 8990-13810, 16088-16335 and 24964-25210 cm^{-1} assigned to $^3A_{2g}$ (F)$\rightarrow {}^3T_{2g}$ (F) $^3A_{2g}$ (F)$\rightarrow {}^3T_{1g}$ (F) and $^3A_{2g}$ (F)$\rightarrow {}^3T_{1g}$ (P) This indicates that the complex of Ni(II) has octahedral geometry.

Table 5.3: Solution Conductivity, Magnetic Susceptibility and Electronic Absorption Spectral Data of Ni(II) Complexes.

Ni(II) Complexes	Molar Conductance Ohm^{-1}cm^{2}mol^{-1}	μ_{eff} (B.M.)	Absorption Maxima cm^{-1} (nm)	
			$^3A_{2g}$ (F)$\rightarrow$ $^3T_{1g}$ (P)	Charge Transfer
B_1	7.82	2.97	23952 (417.5)	30030 (333)
B_2	6.99	3.13	25575 (391)	30628 (326.5)

5.4. Copper (II) Complexes:

Theoretical Considerations:

The Cu(II) ion has d^9 configuration and is most susceptible to John-Teller distortion[98-99] in the both octahedral and tetrahedral environment. In addition to distorted octahedral complexes, some other stereochemistry of Cu(II) is known, of which the most important is distorted tetrahedral, square planar complexes, A wide range of penta-coordinated Cu(II) complexes of both trigonal bipyramidal and square pyramidal types have been reported.

The observed magnetic moment values of octahedral Cu(II) complexes fall in the range 1.8 to 2.1 B.M. The tetrahedral Cu(II) complexes show magnetic moment values in the range 1.9-2.0 B.M. The magnetic moment is expected to be around 1.9 B.M. at room temperature for octahedral or tetrahedral Cu(II) complexes due to a larger separation of interaction term i.e. spin-orbital coupling constant. The theoretically expected magnetic moment value is 2.2 for $^2T_{2g}$ ground term in Cu(II) complexes Figg has suggested normal μ_{eff} value greater than 1.9 B.M. for tetrahedral and less than 1.9 B.M. for square planar or pseudo-octahedral Cu(II) complexes.

Most of the Cu(II) complexes are green or blue because of the single broad absorption band in the region 1000-16000 cm^{-1}. A detailed interpretation of the electronic spectra of Cu(II) complexes is rather complicated because of distorted octahedral or square planar any, distortions are expected to have only one transition $^2E_g \rightarrow ^2T_{2g}$. Due to the Jahn-Teller effect, The system becomes unstable resulting in the further splitting of E_g and T_{2g} into B_{1g}, A_{1g} and B_{2g} level respectively. A broad band with several components is observed at 16000 cm^{-1} in the visible region and often broad tail into the near infrared region. This is due to John-Teller distortion of regular octahedral to tetrahedral geometry.[100] As the distortion of tetrahedral geometry towards square planar continues, the band moves towards higher energy.[101]

Regular tetrahedral Cu(II) complexes give a single band in near-infrared region without any absorption in the region without 1000 to 20000 cm^{-1} corresponding to $^2T_{2g} \rightarrow ^2E_g$ However, low energy bands are observed in the visible region at 14700 cm^{-1} at ($^2B_{1g} \rightarrow ^2A_{1g}$) and 17900$cm^{-1}$ ($2B_{1g} \rightarrow ^2E_g$).[102]

Cu(II) having one unpaired electron, in this case following transitions are possible.

For weakly distorted

$d(x^2\text{-}y^2) \rightarrow d(xy)$	$^2A_g \rightarrow ^1A_{1g}$.
$d(x^2\text{-}y^2) \rightarrow d(z)^2$	$^2B_{1g} \rightarrow ^2B_{2g}$.
$d(x^2\text{-}y^2) \rightarrow d(yz, xz)$	$^2B_{1g} \rightarrow ^2E_g$

For strongly distorted

$d(x^2\text{-}y^2) \rightarrow d(z)^2$	$^2B_{1g} \rightarrow ^2B_{2g}$.
$d(x^2\text{-}y^2) \rightarrow d(yz, xz)$	$^2B_{1g} \rightarrow ^2E_g$
$d(x^2\text{-}y^2) \rightarrow d(xy)$	$^2B_{1g} \rightarrow ^2A_{1g}$.

Thus, the three band due to these transitions are expected to appear in the visible and near IR region and are very close together. Octahedral coordination around metal ions is also supported by the number and position of electronic spectral bands.[103-104]

Elemental Analysis and Solution Conductivity:

All the Cu(II) complexes synthesized are green in color. They are stable to air and moisture. They are insoluble in water but soluble in DMSO and DMF. Most of them are decomposed below 300°C. The observed solution conductivities of Cu(II) complex are presented in Table 5.4 The low conductivity values in DMSO reveal their non-electrolytic nature. The elemental analysis (Table-2.16 & 2.10) shows the 1:2 metal to ligand ratio in all Cu(II) complexes.

Magnetic Susceptibility and Electronic Absorption Spectra of Cu(II) Complexes:

The observed magnetic moment values of Cu(II) complexes in the present investigation are found to be in the range 1.81 to 2.2 B.M. at room temperature corresponding to one unpaired electron. Magnetic moments are slightly higher than the spin only value (1.73 B.M). In the present investigation, Cu(II) show UV transition band in the range 24154 to 26560 cm^{-1} which is attributed to $^2B_{1g} \rightarrow ^2E_g$ and charge transfer band observed in the range 35026-39370 cm^{-1} indicating distorted octahedral geometry around the Cu(II) ion.[105-106] The electronic spectra of these complexes are given in Fig 5.1 to 5.2 and their spectral data are presented in Table 5.4.

Reddy Vidyavati et al.[97] reported that the Cu(II) complexes exhibit three bands in the region 11330, 17300 and 19500-21000 cm^{-1} are of equal energy and giving rise to single broad band which may be assigned to transitions $^2B_{1g} \rightarrow ^2A_{1g}$, $^2B_{1g} \rightarrow ^2B_{2g}$ and $^2B_{1g} \rightarrow ^2E_g$ respectively the broadness of the bond is due to the ligand field and the Jhon-Teller effect, this observation favor the octahedral geometry around the Cu(II) ion.

Table 5.4 : Solution Conductivity, Magnetic Susceptibility and Electronic Absorption Spectral Data of Cu(II) Complexes.

Cu(II) Complex	Molar Conductance $Ohm^{-1}cm^2mol^{-1}$	μ_{eff} (B.M.)	Absorption Maxima cm^{-1} (nm)	
			$^2B_{1g} \rightarrow ^2E_g$	Charge Transfer
B_1	7.2	2.01	24,154(414)	42016 (238)
B_2	6.8	1.98	26,809(373)	41666 (240)

Safana Ahmed Farhan reported Cu(II) complexes of flavonoids derived from morin. In these spectra, morin has two maxima bands the maximum absorption bands is shifted to 226 nm (44247cm^{-1}) and weak band shifted to 292 nm (34246 cm^{-1}) with respect to the free morin resulting observation shows that morin-copper complex has octahedral geometry around the Cu(II) ion.

Mahmoud Najim et al.[107] reported that the green solution of copper complex in DMSO exhibited a broad peak at 835 nm that is due to the resolute transitions of

$^2B_{1g}\rightarrow{}^2B_{2g}$ and $^2B_{1g}\rightarrow{}^2E_g$ thereby supports the distorted octahedral geometry around the Cu(II) ion.

Faize-Beg and et al.[92] reported electronic spectrum of Cu(chrysin)$_2$ complex and it gives d-d transition absorption band at 672.5 nm (14869 cm^{-1}). The observation shows that chrysin-copper complex has octahedral geometry around the Cu(II) ion.

5.5. Zinc (II) Complexes:

Zn(II) has d^{10} system and should be diamagnetic therefore d-d transitions are spin forbidden.[108] The diamagnetic Zn(II) complexes did not show d-d band and its spectrum is dominated only by a charge transfer. The spectra of the Zn(II) complexes exhibited similar behavior to these ligands suggesting square planar geometry.

Elemental Analysis and Solution Conductivity:

All the Zn(II) complexes synthesized are yellow, yellowish brown, white in color except A$_5$Zn complex which is orange in color. They are stable to air and moisture. They are insoluble in water but soluble in DMSO and DMF. The complexes decompose at a high temperature. The observed solution conductivities of Zn(II) complex are presented in Table 5.5. The low conductivity values in DMSO reveal their non-electrolytic nature. The elemental analysis (Table-2.17 & 2.11) shows that the 1:2 metal to ligand ratio in all Zn(II) complexes.

Magnetic Susceptibility and Electronic Absorption Spectra of Zn(II) Complexes:

All the Zn(II) complexes in present investigation are diamagnetic in nature indicating square planar geometry around the Zn(II) metal ion. In the present investigation, d-d transition bands are not observed. Only charge transfer bands are observed in the range of 24875 to 32051 cm^{-1}.

A. Xavier et al.[109] reported mononuclear Zn(II) metal complexes The absorption spectra for zinc complex shows intense band about (291-345 nm) which may be related to ligand field. The absorption peak in the range (315-360 nm) for these complexes is due to the charge transfer since it belongs to d^{10} configuration and does not have d-d transition. The zinc complex shows band at 316 nm assigned to intra-ligand charge transfer transition which is diamagnetic nature and square planar geometry.

Table 5.5 : Solution Conductivity, Magnetic Susceptibility and Electronic Absorption Spectral Data of Zn(II) Complexes.

Zn(II) Complexes	Molar Conductance Ohm^{-1}cm^{2}mol^{-1}	μ_{eff} (B.M.)	Absorption Maxima (LMCT) cm^{-1} Charge transfer.
B$_1$	9.65	Diamagnetic	26809(373)
B$_2$	10.8	Diamagnetic	27027(370)

On the basis of inferences drawn from literature and interpretation of data, element analysis, magnetic and electronic absorption spectral study of complexes, square planar geometry may be proposed to Zn(II) complexes.

5.6. Cadmium (II) Complexes:

Cd(II) has d^{10} system and should be diamagnetic therefore d-d transitions are spin forbidden. The diamagnetic Cd(II) complexes did not show d-d band and its spectrum is dominated only by a charge transfer. The charge transfer band at 319-473 nm was assigned possible in an octahedral environment.

The spectra of the Cd(II) complexes exhibited similar behavior to these ligands suggesting octahedral geometry.

Elemental Analysis and Solution Conductivity:

All the Cd(II) complexes synthesized are yellow in color except A_2Zn complex which is orange in color. They are stable to air and moisture. They are insoluble in water but soluble in DMSO and DMF. The complexes decompose at a high temperature. The observed solution conductivities of Cd(II) complex are presented in Table 5.6. The low conductivity values in DMSO reveal their non-electrolytic nature. The elemental analysis (Table-2.18 & 2.12) shows that the 1:2 metal to ligand ratio in all Cd(II) complexes.

Magnetic Susceptibility and Electronic Absorption Spectra of Cd(II) Complexes:

All the Cd(II) complexes in present investigation are diamagnetic in nature indicating octahedral geometry around the Cd(II) metal ion. In the present investigation, d-d transition bands are not observed. Only charge transfer bands are observed in the range of 21141 to 31347 cm^{-1}.

S.A. Kumari et al.[110] shows the cadmium complexes cadmium metal complex have octahedral geometry.

J. Lachman et al.[111] reported the effect of cadmium on flavonoids content in young barly plants. They measured the UV spectrum of cadmium-flavonoid metal complex, it gives a sharp band at 25316 cm^{-1}.

Table 5.6 :Solution Conductivity, Magnetic Susceptibility and Electronic Absorption Spectral Data of Cd(II) Complexes.

Cd(II) Complexes	**Molar Conductance $Ohm^{-1}cm^2mol^{-1}$**	**μ_{eff} (B.M.)**	**Absorption Maxima (LMCT)cm^{-1}** Charge transfer.
B_1	13.6	Diamagnetic	25188(397)
B_2	13.8	Diamagnetic	25000(400)

On the basic of inferences drawn from literature and interpretation of data, elemental analysis, magnetic and electronic absorption spectral study of complexes, octahedral geometry may be proposed to Cd(II) complexes.

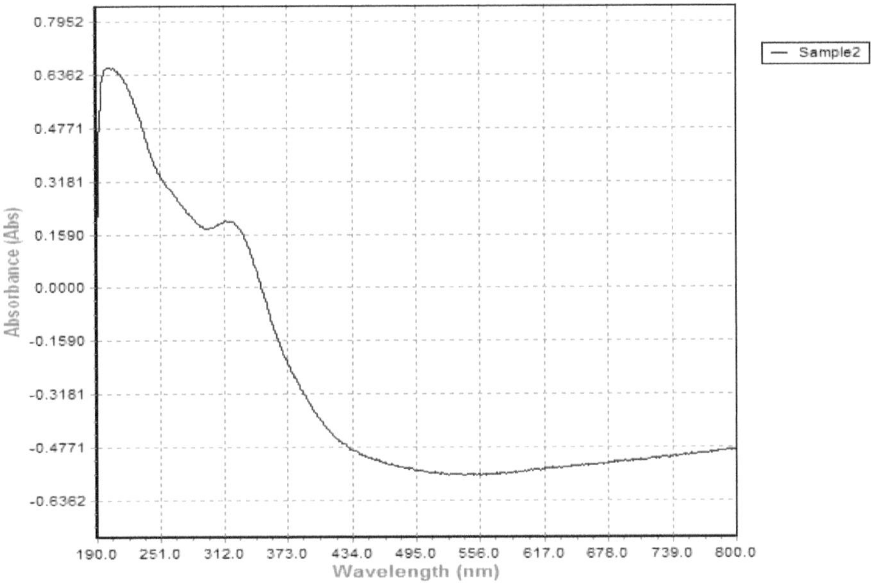

Fig. 5.1: Electronic Absorption Spectrum of Cu(II) Complex of Flavone B_1

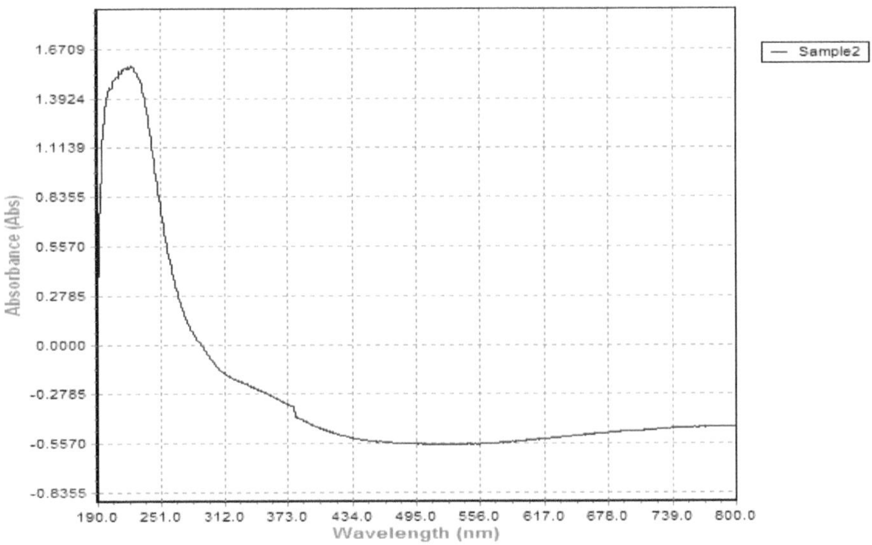

Fig. 5.2: Electronic Absorption Spectrum of Cu(II) Complex of Flavone B_2

CHAPTER 6

Thermal Studies of Metal Complexes

6.1. Theoretical Considerations:

Thermogravimetry (TG) provides a measurement of any weight change associated with a transition. Thermogravimetry is a process in which a substance is decomposed or dehydration in the presence of heat, which causes bonds of the molecules to be broken.[112] The changes in weight are due to the rupture or formation of various physical and chemical bonds at elevated temperature. This technique is widely applied to study the thermal behavior of metal complexes.[113-114] Thermal analysis, including main thermogravimetry (TG) and differential thermal analysis (DTA) are widely used techniques in the qualitative and quantitative evaluation of various compounds.[115]

Thermal properties are important for kinetic studies of thermal decomposition, free energy changes (ΔG), order of reaction (n), frequency factor (Z), entropy of activation (ΔS) and the energy of activation (Ea), which directly govern the factors of thermal stability of any substance. Thermal analysis is a combinational technique of TGA/DTA, using which the physical property of a substance is measured as a function of temperature.

Two alternative methods have been used in the kinetic investigation of thermal decomposition of solids. In the first approach, weight loss-time/temperature measurements are made and in the second approach, the sample is subjected to a controlled rising temperature. The first method is called an isothermal/static method and the second one is called a non-isothermal/dynamic method.

In the static method, determination of isobaric weight change and isothermal weight change is carried whereas in dynamic method, thermogravimetry (TG), derivative thermogravimetry (DTG) and more recent inverted thermogravimetric techniques are used. In non-isothermal TG analysis, more information is extracted from a single TG-curve compared to many curves and hence non-isothermal TG analysis is preferred over isothermal TG analysis.

Thermal analysis instrument measures the temperature of transitions, weight losses in materials, energies of transitions, dimensional changes, dehydration, decomposition, dissociation, phase transformation, heat capacity and purity of the sample viscoelastic properties.

The simultaneous use of TG and DTA is made in the present study of metal complexes with a view to understand stiochiometry, thermal stability, the presence and nature of water molecules. The water in the inorganic compound may be classified as lattice and coordinated water. There is no definite borderline between the two. The former term denotes water molecule trapped in the crystalline lattice either by weak bonds to the anion or by weak ionic bonds to the metal or both, whereas the latter denotes water molecules bonded to the metal through partial covalent bonds or coordinate bonds. According to Freeman and Carroll[116] the water eliminated below 150°C can be considered as lattice water and above 150°C as water coordinated to metal ion.[117]

We have used a dynamic method, 5-10 mg of the sample was treated with increase in temperature by 10°C per minute from 36°C to 680°C in an inert nitrogen atmosphere. Considering percentage loss in weight, thermal decomposition behavior of Mn(II), Co(II), Ni(II), Cu(II), Zn(II) and Cd(II) complexes of some selected ligands are discussed on the basis of their simultaneous TG-DTA curves as follows. All the synthesized compounds show a multistage decomposition pattern.

6.2. Thermal Decomposition Study of Metal Complexes:

Cu (II) Complex :

The thermograms of Cu(II) complex of flavone B_2 are presented in (Fig.6.1). These curves reveal the presence of lattice water as well as coordinated water in the complex.

In Cu(II) complex of B_2 (Fig. 6.1) It is observed that, In the first step weight loss at 89.56°C indicating presence of lattice water with weight loss 8.31% (calc. wt. loss 7.50%). The second loss due to coordinated water molecule liberated in the complex, takes place at 135.29°C with weight loss of 6.95% (calc. wt. loss 7.40%). In the third step loss of 8.97% (calc. wt. loss 9.00%) to the temperature in the range 250-300°C. This is supported by endothermic peak in DTA at 271.92°C due to the fragment of non-coordinated part of ligand. The complex continues to decompose in a fourth step up to 360°C with a weight loss of 8.97% (calc. wt. loss 8.03%) which is confirmed by a sharp endothermic peak in DTA curve at 343.23°C may be due to the decomposition of a fragment of coordinated part of the complex. The last step covering the slow step decomposition of the reaction interval 360-450°C with 3.75% mass reduction corresponds to the decomposition of a fragment of coordinated part of the complex and above which the residue attains almost constant weight due to the formation of CuO as a final product. Thermograms of Cu complexes indicate the presence of coordinated as well as lattice water.

Cd (II) Complex:

The thermograms of Cd(II) complex of flavone B_1 are presented in (Fig.6.2). These curves show the presence of lattice water in the complex.

Cd(II) complex of B_1 (Fig. 6.2) shows two-step decomposition. In first step loss of 3.03% (calc. wt. loss 4.19%) observed from 50-140°C indicating the presence of lattice water molecule which is confirmed by an endothermic peak in DTA at 66.00°C. The second step decomposition observed within 190-340°C with a weight loss of 13.50%. This may be due to the loss of two fragment of furan ring of the chalcone of metal complex. Beyond that TG curve attains a constant level corresponding to their metal oxide that is CdO.

6.3. Thermal Decomposition and Kinetic Parameters of Metal Complexes:

Akahira[118] first introduced that decomposition and kinetic studies of thermal reactions are useful in determining thermodynamic and kinetic parameters like free energy, entropy change, activation energy, pre-exponential factor.

Since then so many methods and equations have been proposed for thermal decomposition and kinetic analysis of reaction. Thermal decomposition studies of materials are useful in predicting thermal stability.

Thermal decomposition kinetics was studied by analytical data. Thermal decompositions kinetics was first studied by Flynn,[119] latter in 1958, Freeman and Carroll published their method of kinetic analysis of thermoanalytical data. Various methods develop for kinetic analysis. Some important methods are Horowitz-Metzger,[120] Coats-Redfern,[121] Sharp-Wentworth,[122] Fuoss,[123] Piloyan-Novikova,[124] MacCallum-Tanner,[125] Dharwadker-Karkhanwal.[126] These methods have been applied depending upon the needs and conditions for kinetic analysis of thermal decomposition of solid substances on the basis of non-isothermal concept.[127-129]

The Horowitz-Metzger method has been used for finding the kinetic parameters for non- isothermal decomposition of Mn(II), Co(II) Ni(II), Cu(II) Zn(II) and Cd(II) complexes of flavones in the present work. By this method, thermal decomposition reaction of many coordination complexes of transition metals has been reported. Coats-Redfern method has also been used for finding the kinetic parameters for non-isothermal decomposition reaction of these complexes. Thermal decomposition reaction of many coordination compounds of higher oxidation states has been studied by this method.[130-131]

Horowitz-Metzger Method:

The Horowitz-Metzger method is illustrative of the approximation methods. The equation used for determining the kinetic and thermodynamic parameters for non-isothermal de composition of complexes by this method is given below,[132]

$$\log\left(\frac{1-(1-\alpha)^{1-n}}{(1-n)}\right)=\frac{E_a\theta}{2.303RT_s^2}+\log\frac{ZRT_s^2}{E\beta}-\frac{E}{2.303RT_s^2}$$

Where,

α= Fraction decomposed	β= Heating rate (10°c/min)
Ts= Temperature at half Wt. Loss.	K= Boltzmann constant.
n= Order of reaction	R= Molar gas constant
Z= Frequency factor	h= Plank's constant.

The left hand side of the above equation was plotted against θ ($T-T_s$). By using different values of order of reaction (n) ranging from 0 to 4 in increment of 0.01, a straight line was obtained by the method of least squares. The highest value of 'r', the correlation coefficient gave the correct value of 'n'. From the slope and the intercept, the energy of activation and pre-exponential factor (frequency factor) were calculated. The value of ΔS and ΔG were determined using E and Z values. The equation used for calculating entropy change (ΔS) is given below,

$$\Delta S = 2.303 R \log \frac{Z\ h}{k\ T_s}$$

Where,

K= Boltzmann constant,

Ts= Peak temperature.

R= Molar gas constant.

β= Rate of heating 10°c/min.

h= Plank's constant

Coats-Redfern Method:

The method of Coats-Redfern is illustrative of the approximation method for reaction order. The weighted least square method (LSM) was used for obtaining best fit linear plots by applying the data of various equations and thus the kinetic parameters were evaluated. Coats-Redfern derived the relation for such reaction by the following equations.

$$\log[1-(1-\alpha)^{1-n}/(1-n)] = E_a\theta/2.303RTs^2$$

or

$$\log 1 - \frac{(1-\alpha)^{1-n}}{(1-n)} = \frac{Ea\theta}{2.303RTs^2}$$

Where,

α= Fraction of weight loss

T= Temperature (°K)

Z= Pre-exponential factor,

q= Heating rate

R= Molar gas constant

E_a= Activation energy

n= Order of reaction (n≠1)

The fraction of weight loss (α) and the corresponding $(1-\alpha)^{1-n}$ have been calculated from TG curves at different temperatures.

$$\alpha = \frac{(Wt - W_f)}{(W_o - W_f)}$$

By plotting a graph between T and (1-α), the values of slope, intercept and energy of activation were obtained from plots. The frequency factor (Z) was calculated by,

$$\text{Intercept} = \log\left(\frac{ZR}{\beta Ea}\right) \quad ...(1)$$

and the entropy of activation (ΔS) was calculated by,

$\Delta S = 2.303 \log$ .... (2)

Where,

K= Boltzmann constant

h= Plank's constant

β= Rate of heating

R= Molar gas constant

T_M = Peak temperature

The thermal kinetic parameters ΔS, Ea and Z for non-isothermal decomposition of complexes have been calculated by Coats-Redfern method from TG-DTA curves (Fig 6.1 to 6.2) and are presented in Table 6.1

Generally, with decreasing value of ΔE, the value of Z increases, and a higher value of activation energy suggest higher stability.[85] In the present complexes, the value of E_a decrease with the increasing value of (Z) i.e. frequency factor indicating that the activated complexes have more ordered or more rigid structure than the reactants or intermediate and that the reactions are slower than normal. The negative values of the entropy of activation (ΔS) indicate that the metal complexes are thermally stable. ΔG is positive for the complexes revealing that the free energy of the final residue is higher than that of the initial complex, and all decomposition steps are non-spontaneous processes. Also, the value of free energy of activation, ΔG increases significantly for the subsequent decomposition stages of a given complex.[133]

Table 6.1 : Thermodynamic and Kinetic Parameters

Metal complex	Method	Step	Decomp. Temp.	Order of Reaction	Ea(KJ mol^{-1})	ΔS(KJ mol^{-1})	ΔG(KJ mol^{-1})	Z (S^{-1})	Correlation Coefficient (r)
B_2Cu	H-M	I	350	0.3	26.94	-160.14	38.34	53456.11	0.989
	C-R				11.23	-81.46	17.03	6.86E+08	0.999
	H-M	II	570	0.3	20.20	-164.28	31.91	32548.02	0.987
	C-R				21.66	-94.03	28.36	151620017.3	0.999
B_1Cd	H-M	I	300	1.1	13.42	-168.08	25.38	20557.36	0.999
	C-R				26.35	-108.07	34.04	27984166	0.999
	H-M	II	600	1.1	94.27	-120.46	92.18	6317297.95	0.990
	C-R				91.08	-77.04	82.08	3.79528E+72	0.998

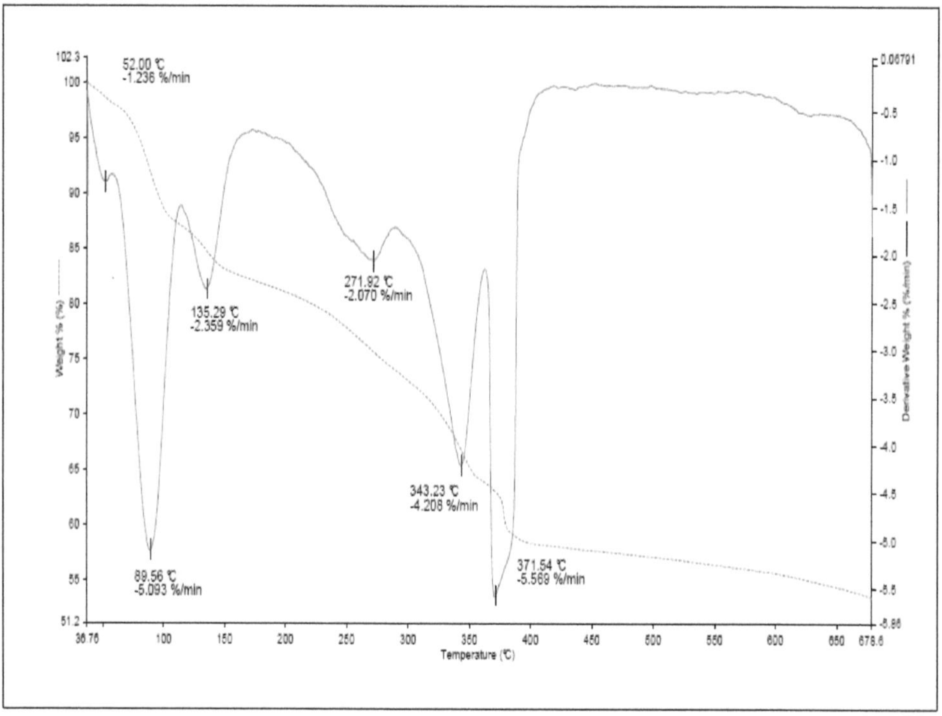

Fig. 6.1: TG-DTA Curve of Cu(II) Complex of flavone B_2

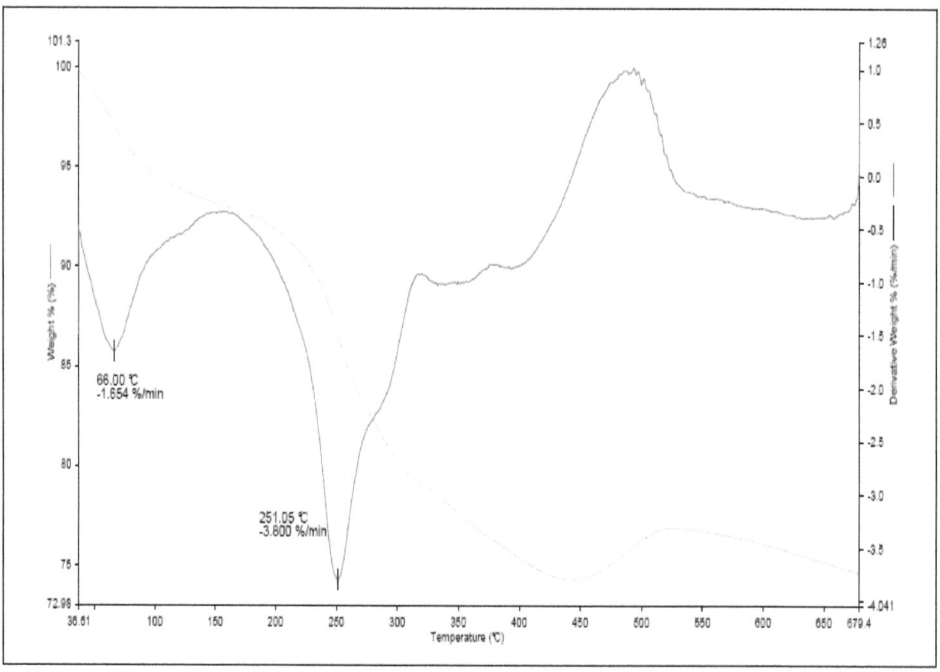

Fig. 6.2: TG-DTA Curve of Cd(II) Complex of flavoneB_1

CHAPTER 7

X-Ray Diffraction Spectral Studies of Metal Complexes

7.1. Theoretical Consideration:

X-Ray crystallography is a technique of determining the arrangement of atoms within a crystal. X-ray Diffraction is a tool for the investigation of the fine structure of matter. This technique had its beginning in Von Laue's discovery in 1912 that, crystal diffract X-rays, the manner of the diffraction revealing the structure of the crystal using this technique, a clear cut idea can be obtained about the arrangement and spacing of atoms in crystalline materials.[134-135] Diffraction pattern can be described in terms of three dimensional array called lattice points. The simplest array of points from which a crystal can be created is called a unit cell. It is possible to diffract X-ray by means of crystals because the wavelength of X-ray is of about the same order as the inter-atomic distances in a crystal. The X-ray interacts with the inner most electronic cloud in the atoms and provides information about internal arrangement of the atoms in the crystal.[136-137] The intensity of the diffracted beam is related to the structure of the crystal. The cation distribution in the sample can be determined by interpreting the XRD patterns and it is a direct tool for assigning the crystal pattern for the complexes.

X-ray method used in the determination of the structure of metal complexes has increased immensely during the last decade. Depending on these values of axial length and axial angles, the crystals are classified as Cubic, Tetragonal, Orthorhombic, Trigonal, Hexagonal, Monoclinic and Triclinic structures. By the X-ray crystallographic studies the determination of unit cell dimensions, lattice type of crystal, the interplanar spacing of lattice planes and Miller indices of the reflection planes are possible.[138-140]

The X-ray powder diffractogram of representative metal complexes synthesized were used for crystal structure characterization and determination of lattice dimensions. The observed data of compounds under investigation were compared with other literature data having analogous cells and subsequent index to similar geometry.

The X-ray diffractograms of representative metal complexes were scanned in the range 0-60° at λ =1.543 Å. The diffractograms and associated data depict the 2θ value for each peak, relative intensity and observed interplanar spacings (d-values). The information obtained from the diffractogram and associated data were used for indexing the pattern and to find out the unit cell dimensions and space group. The cell data of metal complexes along with the powder diffraction data are given in Table 7.1. Various methods have been used for indexing the powder data. The position of each reflection with the intensity was recorded. The interplanar spacing d_{hkl} were calculated from 2θ values using the relation d= nλ/2sinθ.

The computer program, used for indexing data was Powder-X. The preliminary data in the form of 2θ and intensities fed to the computer program and all differences (d-observed) are calculated as required. Values are arranged in the decreasing order. The values can be readily compared and checked with the observed data.

In this program all the essential features of X-ray program are presented and in addition it calculates the deviation in lattice parameter a, b, c in Å and α, β and γ in degree and minutes with better combination of h, k, l values untill the final deviation is within the permissible limit of 2%. The refined values also give the volume of the unit cell.

When the system of a compound is unknown, the observed data is first tested with the isometric chart and Hul-Devy's Curve for a tetragonal and hexagonal system. If the tests are proved to be negative then the data is subjected to program bar and an attempt is made to index it an orthorhombic system. The failure of all these tests is an indication for the existence of lower symmetry and then data can be indexed on the either monoclinic or triclinic system. The precise lattice parameters and the deviation are obtained from powder -X program.

Density and Space Group Determination:

A density determination becomes essential for assigning the space group and the indexing crystal system. Therefore densities of the metal complexes used for XRD study were determined using specific gravity bottle by the principle of Archinedes.[141] Using these observed values of densities and other data obtained from indexing of XRD pattern the formula factor Z was calculated. The formula factor Z was rounded up to the nearest whole number and thus the density was recalculated and was taken as calculated densities. The closeness of observed and calculated values of density on the basis of unit cell volume was used to judge the assignment. The percentage of porosity was calculated from the observed and calculated densities. By using the literature value for definite crystal system, the probable space group was assigned to each of the complexes.

The program calculates various parameters and standard deviations in the calculated lattice parameters a, b, c in Å and α, β and γ in degrees and minutes. The refinement of the lattice parameters is continued with a better combination of hkl values until the final deviation is within the permissible limit of 2%. The refined values also give the volume of the unit cell.

The density ϱ in gcm^{-3} of the compound is related to the formula factor Z and volume of the unit cell V in $Å^3$as per the following expression.

$$\rho = \frac{ZM}{VN_A}$$

Where,

Z= Number of formula units present within unit cell.

M= Molecular formula weight of the compound.

V= Volume of unit cell and

N_A= Avogadro's number.

Results and Discussion:

The X-ray powder diffractograms of the metal complexes were used for the structural characterization and determination of lattice dimensions. The observed data of complexes under investigation were compared with other literature data having analogous cells and subsequently indexed to similar geometry. The original diffraction patterns of complex presented in (Fig 7.1) respectively.

7.2. X-Ray Diffraction Studies of Metal Complexes:

Manganese (II) Complex:

The Mn(II) complex of flavone B_1 complex is selected for X-ray powder diffraction studies. The X-ray diffraction pattern of the complexes is shown in (Fig 7.1). The unit cell data, lattice parameters and the data obtained after indexing the powder pattern is presented in Table 7.1. The standard deviation observed was within the permissible range. The observed density for Mn(II) complex of ligand B_1 is 0.8615 gcm^{-3} while calculated density from Z value and unit cell volume for complex are 0.8609 gcm^{-3}. The standard deviation is 0.19 within the permissible range of 2 % closeness in the two density values indicates the perfection in indexing the X-ray diffraction pattern. The crystal system of Mn(II) complex of B_1 ligand have unit cell volume 1546.5 $Å^3$. The Mn(II) complex of B_1 contain one molecule per unit cell having probable space group P2/m.

Thenmozhi Muthurajan et al.[142] performed the X-ray diffraction study of flavonoid complexes of quercetin. The compound was crystallized in a monoclinic crystal system with space group P2/m. The lattice parameters reported are a= 8.946 Å, b= 3.787 Å, c=7.646 Å, β= 104.5°.

Omar H. Al-Obaidi et al.[143] synthesized and characterized Mn(II) complexes of flavonoids. Reheb Abdul Mahdi Al-Hassani et al.[144] synthesized and characterized Mn(II) complexes of flavonoids. The X-ray powder diffractogram suggests that the present Mn(II) complexes of B_1 ligands under investigation crystallizes in a monoclinic crystal system with 1 atom per unit cell and P2/m.

Table 7.1 : Indexed X-ray Diffraction Data of Mn(II) Complex of Flavone B_1.

Peak No	2θ observed	2θ calculated	d observed	d calculated	Miller Indices of Planes			Relative Intensities (%)
					h	K	L	
1	16.413	16.345	5.39657	5.67843	0	0	1	100
2	18.223	18.324	4.86422	4.85438	0	-1	0	62.63
3	18.238	18.248	4.86032	4.54789	-1	0	0	14.80

Peak No	2θ observed	2θ calculated	d observed	d calculated	Miller Indices of Planes			Relative Intensities (%)
					h	K	L	
4	23.901	23.894	3.72000	3.45873	-2	-1	0	19.18
5	23.932	23.432	3.71526	3.56739	-1	0	1	13.77
6	23.965	23.119	3.71021	3.54893	1	-1	0	5.9
7	24.083	24.345	3.69229	3.76593	1	-1	1	8.37
8	25.145	25.438	3.53828	3.86739	0	1	1	4.5
9	33.323	33.659	2.68664	2.56789	1	1	1	7.75
10	36.929	36.547	2.43231	2.47632	0	-2	0	23.71
11	46.135	46.432	1.96599	1.94678	-1	-2	1	13.19
12	48.158	48.321	1.88016	1.45680	1	2	1	4.9
13	49.057	49.489	1.85550	1.85409	-2	2	0	2.31

Unit cell data and crystal lattice parameters

a (Å) = 7.9163 Volume (V) = 1546.5 $Å^3$

b (Å) = 4.9165 Density (obs.) = 0.8615 gcm^{-3}

c (Å) = 8.4089 Density (cal.) = 0.8609 gcm^{-3}

$\alpha = 90^\circ$ Z = 1

$\beta = 102^\circ$ Crystal system= Monoclinic

$\gamma = 90^\circ$ Space group = P2/m

Standard deviation (%) = 0.042

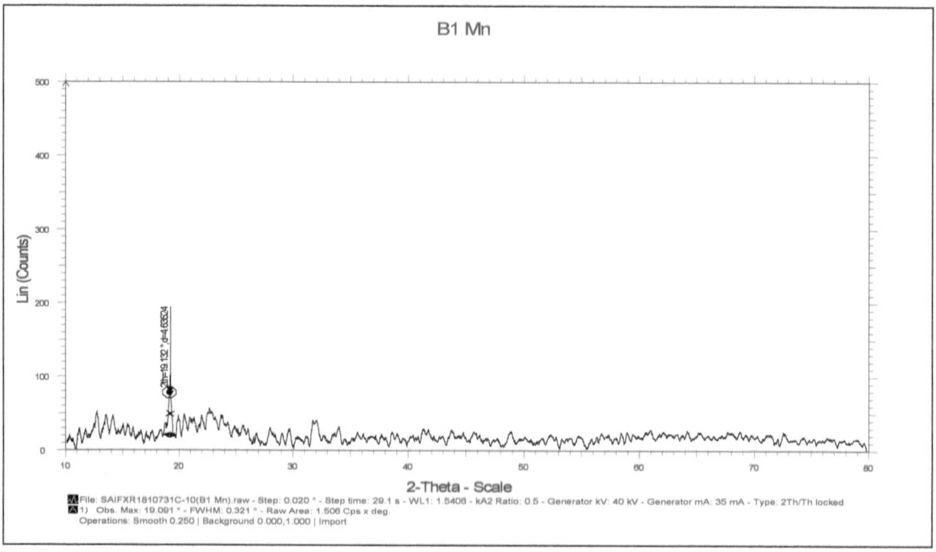

Fig. 7.1: X-ray Diffractogram of Mn(II) Complex of B_1.

CHAPTER 8

Biological Studies of Flavones and their Metal Complexes

8.1. Introduction to Antifungal Activity:

In the treatment of diseases, inhibitory chemicals employed to kill micro-organisms or prevent their growth, are called the anti-microbial activity. The biotic component of the world includes a vast variety of living organism. The biotic component which includes all the living organisms belonging to five kingdoms.

The fungi represent one of the most important kingdoms of the living organism. All fungi are eukaryotic, which means that they do have a nucleus. Fungi have cell walls, but they do not have chloroplast. Fungi do not have chlorophyll and do not make their own food. Fungi do not have specialized tissues and organs such as leaves and roots. Fungi grow best in warm, damp areas, such as tropical forests or between toes.

Fungi have eukaryotic cellular organization and possess a heterotrophic mode of nutrition. They can't synthesis their own organic food through photosynthesis or chemosynthesis and instead obtain it from other sources.

Fungi are a type of heterotroph, which is an organism that cannot make its own food and must eat other organisms. Hyphae release enzymes that help externally digest food that the fungus absorbs from another organism. Most fungi are decomposers or saprophytes (SAP ruh fites), meaning they get food by absorbing dead or decaying tissues of other organisms.

The fungi may be unicellular or multicellular with uni, bi or multinucleate cell. The heterotrophic mode of nutrition in fungi may be saprophytic, parasitic, having a definite cell wall which we find in advance plants. The cell wall composition is not similar in all the fungi. In some fungi, cellulose is the chief cell wall constituent. The external factors such as the composition of the medium used for fungal cultivation, pH value and temperature influence the composition of fungal cell walls.[145-146] Fungi reproduce both asexually and sexually. Reproduction of fungi takes place through the spores. Spores are specialized cells one or more celled, set apart for reproduction. They arise as buds on specialized hyphae or may be born in special. The spore's fruits (endogeneous) on germination, each spore form a new haploid mycelium. Fungi also show a vegetative method of reproduction.

The vegetative multiplication may occur by fragmentation, fission and budding.

The fungi helps in recycling nutrient and play a significant role in some fungi to help in improving the industrial production of various substances like gluconic acid, citric acid, alcohol, other organic acids and enzymes. Fungi also used in agriculture, medicine, food and nutrition. The genus *Trichoderma* a well known bio-control agent used in presentday agriculture for controlling a safe and eco-friendly way of plant disease.[147] *Saccharomyces cerevisiae* or Baker's yeast, a single-cell fungus, is used to bake bread and other wheat products.[148]

Some mushrooms used as therapeutics in traditional and folk medicines, such as chinese medicine.[149] *Penicillium chrysogenum* and other fungi produces antibiotics, such as *penicillin* and *cepalosporine*[150] etc. Worldwide use of these antibiotics for the treatment of bacterial diseases, such as leprosy, tuberculosis, syphlis and many others. 'White rot fungi' can degrade insecticides, herbicides, coal tars, creosote and heavy fuels and turn them into carbon dioxide, basic element and water.[151]

Some fungi act as pathogens for plants and animals. They can destroy timber, textile. food and leather. In the present study three fungi *Aspergillus niger* (ATCC16404), *Saccharomyces cerevisiae* (ATCC9763) and *Candida albicans* (ATCC10231)

Aspergillus niger:

Aspergillus niger is a kind of Ascomycetes includes the fungi which are commonly called as black mold or black *Aspergillis*. *A. niger* is a plant deleterious fungi and air born ascomycetes to which *A. niger* belongs are found on paddy crops. The fungus *A. niger* is omnivorous in occurrence. The fungi are always associated with food grains, fruits and vegetables during storage and cause spoilage to these stored products.

A. niger is a fungus of great biotechnological importance as it synthesizes useful product of high commercial value. *Aspergillus* are highly useful in various industrial processess such as fermentation, baking, brewing and industrial manufacturing of organic acids, fats, cheese, enzymes, vitamins and antibiotics because of their ability to secrete digestive enzymes. Using *A. niger,* citric acid and gluconic acid are manufactured commercially on large scale.[152] In Japan, *A. oryzae* is used in fermentation to manufacture sake, an alcoholic beverage from rice starch and to make various fermented food products. *A. niger* is used in the study of many biological processess and can detect traces of a quantities of elements Cu, Fe, Mn and Zn from the unknown sample.

The genus is widely distributed in nature. A huge number of spores of these organisms are suspended in air and responsible for variety of allergies and respiratory disorders in a human being. *A. fumigatus, A. flavus, A. niger* and other species cause Aspergillosis disease. The Aspergillosis of the lungs is much more common in birds. It is also common in cattle, sheep, horse and occurs rarely in a human being.[153] Several species grow on leather, timber, cloth and reduced their commercial value. They are also common contaminants of cultures in bacteriological and mycological laboratories.

Saccharomyces cerevisiae:

Saccharomyces cerevisiae commonly known as baker's yeast is a single-celled eukaryote that is frequently used in scientific research. *S. cerevisiae* is an attractive model organism due to the fact that its genome has been sequenced, its genetics are easily manipulated, and it is very easy to maintain in the laboratory. *Saccharomyces cerevisiae* is a species of yeast. It has been instrumental to winemaking, baking, and brewing since ancient times. It is believed to have been originally isolated from the skin of grapes and on the skins of some dark-colored fruits such as plums; it exists among the waxes of the cuticle. It is the microorganism behind the most common type of fermentation. *S. cerevisiae* cells are round to ovoid, 5-10 μm in diameter. It reproduces by a division process known as budding.[154] The optimum temperature for growth of *S. cerevisiae* is 30-35 °C.[155] The budding yeast *Saccharomyces cerevisiae* is a powerful model organism for studying fundamental aspects of eukaryotic cell biology. *Saccharomyces cerevisiae* has both asexual and sexual reproduction. There are two forms in which yeast cells can survive and grow, haploid and diploid. The haploid cells undergo a simple lifecycle of mitosis and growth, and under conditions of high stress will, in general, die. This is the asexual form of the fungus. The diploid cells similarly undergo a simple lifecycle of mitosis and growth. The rate at which the mitotic cell cycle progresses often differs substantially between haploid and diploid cells.[156] Under conditions of stress, diploid cells can undergo sporulation, entering meiosis and producing four haploid spores, which can subsequently mate.

All strains of *S. cerevisiae* can grow aerobically on glucose, maltose, and trehalose and fail to grow on lactose and cellobiose.

Candida albicans:

Candida albicans is an opportunistic pathogenic yeast[157] that is a common member of the human gut flora. It does not proliferate outside the human body.[1158] It is detected in the gastrointestinal tract and mouth in 40–60% of healthy adults.[159-160]

It is usually a commensal organism, but it can become pathogenic in immunocompromised individuals under a variety of conditions.[160-161] It is one of the few species of the genus *Candida* that causes the human infection candidiasis, which results from an overgrowth of the fungus.[160-161] *C. albicans* is commonly used as a model organism for biology. It is generally referred to as a dimorphic fungus since it grows both as yeast and filamentous cells. *C. albicans* is a polymorphic fungus that can grow either as ovoid-shaped budding yeast, as elongated ellipsoid cells with constrictions at the septa (pseudohyphae) or as parallel-walled true hyphae. Further morphologies include white and opaque cells, formed during switching, and chlamydospores, which are thick-walled spore-like structures. *Candida albicans* is a dimorphic fungus. This means that that *C. albicans* has to different phenotypic forms, an oval-shaped yeast form and a branching hyphal form. *C. albicans* normal habitat is the mucosal membranes of humans and various other mammals including the mouth, gut, vagina, and sometimes the skin. Normally *C. albicans* causes no damage and lives symbiotically with the human or animal host, even helping to breakdown minute amounts of fiber that are eaten in the host's diet. The normal bacterial flora of the gut, mouth, and vaginal mucosa act as a barrier to the overgrowth of fungal infections like *C. albicans*. Loss of this normal flora is one of the main predisposing factors to an infection by *C. albicans*. Common symptoms of an oral *C. albicans* infection include burning pain, altered taste, difficulty swallowing and whitish spots on the gums and tongue. Common symptoms of a vaginal *C. albicans* infection include itching, swelling, and thick and odorous discharge. Vaginal fungal infections are usually associated with pregnancy. It is also estimated that every women in her lifetime will have had a *C. albicans* infection.

Fungal Growth:

When microorganisms are inoculated in a suitable medium and incubated under appropriate conditions, a tremendous increase in the cell mass or a number of cells occurs within a relatively short time. This is called microbial growth. Actually, growth may be defined as an irreversible increase in mass of the whole or part of a living organism by the synthesis of macromolecules.

Factor Affecting on Fungal Growth:

Following factors affect on the fungal growth.

Temperature:

For any specific organism there will be a minimum and maximum temperature which refers to the temperature at which considerable growth occurs. The optimum temperature is the temperature at which the growth rate is high. Temperature is an important environmental factor affecting growth and mycotoxin production by moulds. Fungi are capable of surviving under the full range of temperatures normally experienced in environments in which they live. The temperature range usually reported for fungal growth is broad (10-35°C), with a few species capable of growth below or above this range. Fungi can be divided according to their tolerance to a temperature in psychrotolerant, psychrophilic, mesophilic, thermotolerant and thermophilic fungi.

pH of the Medium :

Enzyme activity is also known to be conditioned by the composition of the medium, although different enzymes have different pH optima for their activity. The general favorable range lies between pH 4 to pH 8. Hydrogen ion concentration in a medium could affect growth either indirectly by its effect on the availability of nutrients or directly by action on the cell surfaces. The acid/alkaline requirement for growth of all yeasts and moulds is quite broad, ranging from pH 3 to above pH 8, with optimum around pH 5, if nutrient requirements are satisfied. In general, *Aspergillus* species are more tolerant to alkaline pH while *Penicillium* species appear to be more tolerant to acidic pH (Wheeler et al., 1991).

Humidity:

The water content of the atmosphere is expressed in terms of relative humidity, the ratio of the water vapor pressure of the gas phase being considered, to that of a saturated atmosphere at the same temperature. It is hence the same ratio as water activity but expressed as a percentage. Relative humidity between 95 to 100 % generally supports efficient growth of most fungi and that below 80-85 % inhibit their growth.

Concentration:

A few works have studied the effect of concentration of essential trace elements for their optimum growth. A concentration of essential trace elements higher than the optimum has been found to be an inhibitor for the growth of different fungi studied by them. To control the harmful effect of fungi and also to test the biological application of the newly synthesized complexes. It was felt wiser to undertake the antifungal study of the prepared compounds. To ward off the harmful effect of these friendly fungi and to search control measures for such effects and also to test the biological applications for the newly synthesized complexes it was felt to undertake the study of the present complexes.

Light:

There are some reports that illumination will increase or more commonly reduce the rate at which fungi spread across an agar surface. Such effects are sometimes due to the photochemical destruction of components of the medium but in other instances, a direct effect on metabolism seems likely. The biosynthesis of pigments, mainly carotenoids, as a consequence of light action has been demonstrated.

8.2. Experimental: Antifungal Activity

The flavones and their metal complexes are screened *in vitro* for antifungal activity. The fungal toxicity of chalcones, flavones and their metal complexes were studied *in vitro* against *Aspergillus niger* ATCC 16404, *Saccharomyces* cerevisiae ATCC 9763, *Candida albicans* ATCC 10231 fungal pathogens at fixed 1% concentration.

The species were collected from Department of Microbiology Dr. Babasaheb Ambedkar Marathwada University Sub Campus Osmanabad.

Experimental Procedure for Antifungal Activity:

The in vitro antifungal activity of all synthesized compounds was assessed by using agar well diffusion method by Mancini et al., (2004) with some modifications. Antifungal activity of synthesized compounds was determined against *Aspergillus niger* ATCC 16404, *Saccharomyces cereviseae* ATCC 9763, *Candida albicans* ATCC 10231 fungal pathogens. Fluconozole was used as an antifungal standard reference compound. Synthesized compounds were dissolved in DMSO at a concentration of 1mg/ml. Each fungus was inoculated into sterile Nutrient broth medium and kept at 37 ^{0}C for 24 hours for developing inoculums, and then this broth was used for the study. 200 μL diluted suspension of each pathogen was inoculated on sterile Mueller Hinton agar plates. Wells were punched in the agar medium. Using Micropipette, 100 μl of each compound solution was put in a separate well. 100 μl of DMSO solution without any compound was also placed in a well to check its activity against the pathogenic culture. All Petri dishes were incubated for 24 hours at 37 ^{0}C. A clear zone around the well was considered as positive results. After complete incubation, the antifungal activity of the synthesized compounds was measured. Zones were measured and recorded by using scales in millimeter (mm) are shown in table no. 8.1-8.2.

Table 8.1 : Antifungal Activity of Flavones

Ligands	**Antifungal Growth**		
	Aspergillus niger **ATCC 16404**	***Saccharomyces cereviseae*** **ATCC 9763**	***Candida albicans*** ATCC10231
	1%	1%	1%
B_1	12	13	12
B_2	10	11	08
Standard	26	24	34

Table 8.2: Antifungal Activity of Flavones and their Metal Complexes

Test Compounds	***Aspergillus niger*** **ATCC 16404**	***Saccharomyces*** **cerevisiae ATCC 9763**	***Candida albicans*** ATCC10231
	1%	1%	1%
B_1	**12**	**13**	**12**
B_1-Mn	--	--	06
B_1-Co	12	10	15
B_1-Ni	--	--	--
B_1-Cu	--	--	--
B_1-Zn	--	11	06
B_1-Cd	14	16	23
B_2	**10**	**11**	**08**
B_2-Mn	--	--	--
B_2-Co	--	--	--
B_2-Ni	--	--	--
B_2-Cu	--	--	--

Test Compounds	*Aspergillus niger* ATCC 16404	*Saccharomyces* **cerevisiae ATCC 9763**	*Candida albicans* ATCC10231
	1%	1%	1%
B_2-Zn	10	09	11
B_2-Cd	11	13	18
Standard	**26**	**24**	**29**

8.3. Results and Discussion: Antifungal Activity

The flavones and their metal complexes are screened *in vitro* for antifungal activity. The fungal toxicity of flavones and their metal complexes are studied *in vitro* against *Aspergillus niger* ATCC 16404, *Saccharomyces cerevisiae* ATCC 9763, *Candida albicans* ATCC10231 fungal pathogens at fixed 1% concentration.

Fluconozole was used as an antifungal standard reference compound. The ligand B_1 and B_2 show moderate to good antifungal activity against *Aspergillus niger* ATCC 16404, *Saccharomyces cerevisiae* ATCC 9763, *Candida albicans* ATCC10231 some Ligands do not show antifungal activity but their complexes show appreciable activity. Antifungal activity of complexes increased several times on being coordinated with metal ions.

The Cd(II) complex show more than 70% reduction of fungal growth for all fungi, Cu(II), Co(II), Mn(II), and Zn(II) complexes also show moderate fungal growth reduction and Ni(II) complexes does not show reduction in fungal growth for all fungi.

The fungicidal activity of the metal complexes may be due to the change in structure because of coordination or chelation which tend to make complexes more powerful fungicidal agents, thus inhibiting the growth of fungi.[144] Fungicidal activity in complexes is higher because of the increased lipid solubility of the metal complexes as compared to ligand towards fungal cell membrane.[145-146] This increased lipophilicity enhance the penetration of the complexes into lipid membrane.

These results indicate that the complexes show more activity and the ligands have comparatively less activity against some organisms under identical experimental conditions. This would suggest that the chelation facilitate the ability of complexes to cross a cell membrane and can be explained on the basis of Tweedy's chelation theory and overtone's concept. Overtone concept states that, antimicrobial activity in complexes is higher because of the increased lipid solubility of the metal complexes as compared to ligand towards the cell membrane. The antifungal data reveals that the metal complexes show enhanced activity, most of the metal complexes show better antibacterial activity.

8.4. Introduction to Antibacterial Activity:

The bacteria are microscopic organisms with relatively simple and primitive forms of prokaryotic type. Danish Physician Christian Grams, discovered the differential staining technique known as Gram staining, which differentiates the bacteria into two groups "Gram positive" and "Gram negative", Gram positive bacteria retain the crystal violet and resist decolorization with acetone or alcohol and hence appear deep violet in color; while Gram negative bacteria, which lose the crystal violet, are counter-stained by safranin and hence appear red in color.

Microorganisms are ubiquitous in occurrence. By virtue of their minute body size and simplicity of structures, they reproduce at a rapid rate and bring about physical and chemical changes in the environment and cause beneficial as well as a harmful effect on a human being. Microorganisms are the main cause of any metabolic disturbance occurring in the human body. The disease-causing agents are called pathogens. Pathogenic bacteria cause disease to human, such as cholera, typhoid fever, foodborne illness and tuberculosis. Bacterial diseases are also common in plants, which include leaf spot, fire blight and wilt in plants.

The growth of pathogens and other harmful bacteria is controlled by antibacterial agents, which are essential for the survival of the mankind. The bacterial infection may be treated with antibiotics. The antibiotics are natural as well as synthetic molecules. They just inhibit the growth of bacteria or kill the bacteria. They are produced by micro-organisms or they might be fully or partly prepared by chemical synthesis. They inhibit the growth of micro-organisms in minimal concentrations. Antibiotics may be of microbial origin or purely synthetic or semi-synthetic.

The use of metal complexes as antibacterial agents and the chelates playing the role of certain drugs has been realized in recent decades. Majority of the complexes have biocidal activity.

Bacillus subtilis:

Bacillus subtilis are Gram-positive, rod-shaped bacteria[162] that are naturally found in soil and vegetation. They grow in optimal temperature 25-35°C. They are aerobic. These bacteria are a good model for cellular development[163] and differentiation. *Bacilus subtilis* is a model organism for studying endosperm formation in bacteria.[164] *B. subtilis* strains can act as bio fungicides for benefiting agricultural crops and antibacterial agents.

B. subtilis bacteria are non pathogenic but they can contaminate food, they seldom result in food that is non-acidic, and cause food poisoning, such as *B. licheniformis.*[165] *B. subtilis* is only known to cause disease in several immuno compromised patient and can conversely be used as probiotic in healthy individuals.[166]

Bacillus cereus:

Bacillus cereus has belonged to family Bacillaceae. *Bacillus cereus* is a Gram-positive, rod-shaped, aerobic, facultatively anaerobic, motile, beta-hemolytic bacterium commonly found in soil and food. Some strains are harmful to humans and cause foodborne illness, while other strains can be beneficial as probiotics for animals.[167-168] Bacillus foodborn illnesses occur due to survival of the bacterial endospores when food is improperly cooked.[169]

Bacillus megaterium:

The species was described by De-Bary in 1884 it is called *Bacillus megaterium* and belongs to family bacillaceae. *Bacillus megaterium* is a rod-like, Gram-positive, mainly aerobic spore forming bacterium found in widely diverse habitats. It is used for an intensive study on sporulation and bacteriophages and in the field of biotechnology for its recombinant protein production capacity. With a cell length of up to 4 μm and a diameter of 1.5 μm, *B. megaterium* is amongst the biggest known bacteria.170 *B. megaterium* grows at temperatures from 3 °C to 45 °C, with the optimum around 30 °C.

Micrococcus glutamicum:

It belongs to family Corynebacteriaceae. Corynebacterium glutamicum previously known as Micrococcus glutamicus[171] is a Gram-positive, rod-shaped bacterium that is used industrially for large-scale production of amino acids.[172] While originally identified in a screen for organisms secreting L-glutamate, mutants of C. glutamicum have also been identified that produce various other amino acids.[173]

Staphylococcus aureus :

S. aureus was discovered in Aberdeon Scotland in 1880 by the surgeon, Sir Alexander Ogston in pus from the surgical abscess.[174] *S. aureus* is a Gram-positive bacteria which appears as grape-like cluster when viewed through a microscope and largened round golden-yellow colonies often with B-hemolysis when grown on blood agar plates.[153] The golden appearance is the etymological root of the bacteria's name, aureus means "golden" in Latin.

S. aureus lives on the skin or in the nose of a healthy person that can cause a range of illnesses from minor skin infections like pimples and abscesses to life threatening diseases such as Pneumonia meningitis endocarditis, Toxic shock syndrome (TSS) and septicemia.

Escherichia coli:

E. coli is a Gram-negative bacterium. It is one of the main species of bacteria living in the lower intestine of mammals known as gut flora. The *E. coli* discovered in 1885 by Theodor Escherich, a German pediatrician and bacteriologist. As a result of their adaptation to mammalian intestines, *E. coli* grow best in vivo or at the higher temperatures characteristic of such an environment rather than the cooler temperatures. *E. coli* generally cause intestinal and extraintestinal infections meningitis peritonitis mastitis septicemia and Gram-negative pneumonia *E. coli* cause diarrhea in humans, pigs, sheep, goats, cattle, dogs and horses.[175]

Shigella boydii:

The species is named by the American bacteriologist Mark Frederick Boyd. It belongs to family Enterobacteriaceae. *Shigella boydii* is a Gram-negative bacterium of the genus Shigella. Like other members of the genus, *S. boydii* is a nonmotile, nonspore forming, rod-shaped bacterium which can cause dysentery in humans through fecal-oral contamination.[176] *S. boydii* is the most genetically divergent species of the genus Shigella.[177]

Enterobacter aerogenes:

The Enterobacteriaceae family includes genera Enterobacter The Gram-negative bacteria resides in soil, water, dairy products and inhabits on a natural flora in the gastrointestinal tract of animals as well as humans.

Enterobacter aerogenes causes disease in humans through inadvertent bacteria transfer in hospital settings. Infections commonly attributed to E. aerogenes are respiratory, gastrointestinal, and urinary tract infections, specifically cystits, in addition to wound, bloodstream, and central nervous system infections.

Pseudomonas aeruginosa:

Pseudomonas aeruginosa is a common Gram-negative, rod-shaped bacterium that can cause disease in plants and animals, including humans. It belongs to family Pseudomonadaceae A species of considerable medical importance, *P. aeruginosa* is a multidrug-resistant pathogen recognized for its ubiquity, its intrinsically advanced antibiotic resistance mechanisms, and its association with serious illnesses – hospital-acquired infections such as ventilator-associated pneumonia and various sepsis syndromes. *P. aeruginosa* is a facultative anaerobe, as it is well adapted to proliferate in conditions of partial or total oxygen depletion.

Salmonella abony:

A previous genetic investigation involving a mouse-nonvirulent *salmonella abony* doner and a virulent *S. typhimurium* recipient indicated that two unliked low-virulnce loci determined nonvirulence. The recombinant was smooth and prototrophic. The nonproliferation of the recombinant was determined by one or both of the low virulence loci from the nonvirulent *S. abony* donor.

Salmonella typhi:

S. typhi is a Gram-negative, rod-shaped bacilli. *S. typhi* is pathogenic bacteria. Most human infections are caused by eating food or drinking water that has been contaminated by feces.

Salmonella infection or Salmonellosis is a bacterial disease of the intestinal track. Approximately 1.4 millian Americans are affected with Salmonellosis every year, of about 500 die, according to the Center for Disease control and Prevention. Salmonella is a group of bacteria that cause typhoid fever, food poisoning, gastroenteritis, enteric fever and other illnesses. The name typhoid means "resembling typhus" and comes from the neuropsychiatric symptoms common to typhoid and typhus are distinct diseases and caused by different species of bacteria.[178]

8.5. Literature Survey of Antimicrobial Studies:

Omar H. Al-Obaidi et al.[179] synthesized chalcones derived from 2,6-dihydroxy acetophenone and 4-(dimethylamino) benzaldehyde and its Mn(II), Co(II), Ni(II), Cu(II) complexes. He studied their antimicrobial activity against *E. coli* and *S.aureus* in different concentrations. The comparison value of chalcone with their complexes indicates that the metal complexes exhibited higher antibacterial activity than chalcones. Such increased activity of the complexes can be explained based on Overtones concept.[168]

Heena Kaur et al.[180] synthesized biologically active halogen substituted flavones were synthesized by oxidation cyclization of the corresponding halogen substituted 2-hydroxy chalcone. The newly synthesized compounds were tested for their in vitro antibacterial activity against *E.coli, S.typhi, S.aureus and Bacillus subtillis* and antifungal against, *A. niger, A. flavus, F. moneliforme and P. chrysogenum,* which exhibit moderate to good activity against different strains of bacteria and fungi used.

8.6. Experimental: Antibacterial Activity:

The flavones and their metal complexes are screened *in vitro* for antibacterial activity. The antibacterial activity of flavones and their metal complexes were studied, for evaluating antibacterial activity Gram positive and Gram negative bacterial pathogens were used. *Staphylococcus aureus* ATCC 6538, *Bacillus megaterium* ATCC 2326, *Bacillus subtilis* ATCC 6633 were Gram positive pathogens used in this study. *Escherichia coli* ATCC8739, *Salmonella typhi* ATCC9207, *Shigella boydii* ATCC 12034, *Enterobacter aerogenes* ATCC13048, *Pseudomonas aerogenosa* ATCC9027, *Salmonella abony* NCTC6017 were the Gram-negative pathogens used in this study.

The species were collected from Department of Microbiology Dr. Babasaheb Ambedkar Marathwada University Sub Campus Osmanabad.

Experimental Procedure for Antibacterial Activity:

The in vitro antibacterial activity of all synthesized compounds was assessed by using agar well diffusion method by Mancini et al.,(2004) with some modifications. For evaluating antibacterial activity both Gram-positive and Gram-negative bacteria pathogens were used. *Staphylococcus aureus* ATCC 6538, *Bacillus megaterium* ATCC 2326, *Bacillus subtilis* ATCC 6633 were Gram positive pathogens used in this study. *Escherichia coli* ATCC8739, *Salmonella typhi* ATCC9207, *Shigella boydii* ATCC 12034, *Enterobacter aerogenes* ATCC13048, *Pseudomonas aerogenosa* ATCC9027, *Salmonella abony* NCTC6017 were the Gram negative pathogens used in this study.

Tetracycline was used as an antibacterial standard reference compound. Synthesized compounds were dissolved in DMSO at a concentration of 1mg/ml. Each bacterium was inoculated into sterile Nutrient broth medium and kept at 37 ^{0}C for 24 hours for developing inoculums, and then the broth was used for the study. Using sterile saline, the bacterial suspension was diluted to adjust the turbidity to the 0.5 McFarland standards. 200 μL diluted suspension of each pathogen was inoculated on sterile Mueller Hinton agar plates. Wells were punched in the agar medium. Using Micropipette, 100μl of each compound solution was put in a separate well. 100 μl of DMSO solution without any compound was also placed in a well to check its activity against the pathogenic culture. All Petri dishes were incubated for 24 hour at 37^{0}C.

A clear zone around the well was considered as positive results. After complete incubation, the antibacterial activity of the synthesized compounds was measured. Zones were measured and recorded by using scales in millimeter (mm) are shown in table no. 8.3-8.4.

Table 8.3 : Antibacterial Activity of Flavones

Ligands	**Diameter of inhibition zone (mm)**									
	S.typhi **ATCC9207**	***E.aerogenes*** **ATCC13048**	***B.subtilis*** **ATCC 6633**	**P.aerogenosa ATCC9027**	***S.abony*** **NCTC6017**	***B.megaterium*** **ATCC 2326**	***E.Coli*** **ATCC8739**	***S.aureus*** **ATCC 6538**	***S.boydii*** **ATCC 12034**	***B.cereus*** **ATCC 14579**
	1%	**1%**	**1%**	**1%**	**1%**	**1%**	**1%**	**1%**	**1%**	**1%**
B_1	08	12	10	14	13	14	15	14	13	14
B_2	09	08	15	13	11	12	12	10	15	13
Std	27	33	34	33	30	32	29	25	27	33

Table 8.4 Antibacterial Activity of Ligands and their Metal Complexes

Ligands And Metal complexes	Diameter of Inhibition Zone (mm)									
	S.typhi ATCC9207	*E.aerogenes* ATCC13048	*B.subtilis* ATCC 6633	P.aerogenosa ATCC9027	*S.abony* NCTC6017	*B.megaterium* ATCC 2326	*E.Coli* ATCC8739	*S.aureus* ATCC 6538	*S.boydii* ATCC 12034	*B.cereus* ATCC 14579
	1%	1%	1%	1%	1%	1%	1%	1%	1%	1%
B_1	08	12	10	14	13	14	15	14	13	14
B1Mn	--	--	14	--	08	08	06	--	11	14
B1Co	05	08	08	16	12	10	13	12	12	16
B1Ni	--	--	--	--	--	--	--	--	--	--

Ligands And Metal complexes	Diameter of Inhibition Zone (mm)									
	S.typhi ATCC9207	*E.aerogenes* ATCC13048	*B.subtilis* ATCC 6633	P.aerogenosa ATCC9027	*S.abony* NCTC6017	*B.megaterium* ATCC 2326	*E.Coli* ATCC8739	*S.aureus* ATCC 6538	*S.boydii* ATCC 12034	*B.cereus* ATCC 14579
	1%	1%	1%	1%	1%	1%	1%	1%	1%	1%
B1Cu	--	--	--	--	--	--	--	--	--	--
B1Zn	15	12	08	07	06	11	08	15	--	12
B1Cd	15	08	28	27	26	28	22	22	16	10
B_2	09	08	15	13	11	12	12	10	15	13
B2Mn	--	--	13	12	--	--	--	15	--	--
B2Co	--	--	05	--	--	--	--	--	--	--
B2Ni	--	--	--	--	--	--	--	--	--	--
B2Cu	--	--	12	--	--	--	--	--	--	--
B2Zn	08	14	06	16	14	13	08	15	11	10
B2Cd	--	06	09	20	25	14	18	26	--	21

8.7. Results and Discussion: Antibacterial Activity:

The flavones and their metal complexes are screened *in vitro* for antibacterial activity. The antibacterial activity of flavones and their metal complexes were studied, For evaluating antibacterial activity both Gram-positive and Gram-negative bacteria pathogens were used.

Staphylococcus aureus ATCC 6538, *Bacillus megaterium* ATCC 2326, *Bacillus subtilis* ATCC 6633 were Gram-positive pathogens used in this study. *Escherichia coli* ATCC8739, *Salmonella typhi* ATCC9207, *Shigella boydii* ATCC 12034, *Enterobacter aerogenes* ATCC13048, *Pseudomonas aerogenosa* ATCC9027, *Salmonella abony* NCTC6017 were the Gram-negative pathogens used in this study.

The standard antibacterial drug, tetracycline was used as an antibacterial standard reference compound with 25-33 mm zone of inhibition at 1% concentration. However, the activity of these ligands and complexes is considerably less than that of the standard drug. The relation between structure and activity can be established that the presence of a halogen, hydroxyl group in ligand moiety enhances the microbial activity.

Among these compounds the ligands of flavones (B_1 and B_2) shows moderate antibacterial activity.

The antibacterial data reveals that the metal complexes show enhanced activity in case of Gram-positive and Gram-negative bacteria. Most of the metal complexes show better antibacterial activity. The compounds having antibacterial activity may act either by killing the microbe or inhibiting multiplicity of microbes or blocking their active sites.[181]

In the present study, the metal complexes possess higher antimicrobial activity than the ligands. The higher antimicrobial activity of the metal complexes may be due to the change in structure because of coordination and chelation which tends to make complexes more powerful bacteriostatic agents, thus inhibiting the growth of microorgnism.[182] Antibacterial activity in complexes is higher because of the increased lipid solubility of the metal complexes as compared to ligand towards bacterial cell membrane.[183-184] This increased lipophilicity enhance the penetration of the complexes into lipid membrane and blocking the metal binding site on the enzymes of microorganisms.

CHAPTER 9

Summary and Conclusions

9.1 Summary and Conclusions:

The present work was carried out to understand the structural and biological properties of flavones and their metal complexes. Two flavone ligands are synthesized by using chalcones of 2-furaldehyde and 5-methyl-2-furaldehyde by using DMSO/I2 as an oxidizing agent and have been used for the first time in the synthesis and studies on structural aspects of transition metal complexes. Ligands are in color B1 (Coffee Brown), B2, (Yellowish Brown).

The flavones are soluble in DMF and DMSO. The melting points are in the range of 240-272°C indicating thermal stability at room temperature. All of them are diamagnetic in nature. The UV and IR spectra of these ligands exhibited absorption bands in various regions depending on functional groups present in it. These absorption bands are assigned using proper logic.

After characterization of flavones with respect to all above parameters, we are used them to synthesize metal complexes of Mn(II), Co(II), Ni(II), Cu(II) Zn(II) and Cd(II). Each of these complexes was diagnosed with respect to various analytical, spectral and biological investigations.

The findings are as follows.

1. All these metal complexes coordinate with bidentate flavones with 1:2 metals to ligands stoichiometry.
2. All these metal complexes are insoluble in common organic solvents but soluble in DMF and DMSO.
3. All the metal complexes are non-electrolytic in nature as suggested by low value of their molar conductance.
4. They are sufficiently stable at room temperature, since they do not decompose even at 300 °C except some Ni(II) and Cu(II) complexes.
5. The IR spectral data reveals that two ligands (B_1, B_2) involve enolic hydroxyl group of oxygen and Carbonyl group (-C=O in pyron ring) in flavones are involved in the coordination. This was confirmed by the shift of IR frequency of these groups to lower frequencies in metal complexes.
6. From UV, magnetic moment measurements, color of the complexes and position of d-d transitions suggests octahedral geometry for Mn(II), Co(II),

Ni(II), Cu(II), Cd(II) metal complexes but except Zn(II) complex which show square planar geometry .

7. Magnetic susceptibility measurements at room temperature indicate that the Mn(II), Co(II), Ni(II), Cu(II) complexes are paramagnetic in nature whereas Zn(II) and Cd(II) complexes are diamagnetic in nature.
8. The TG-DTA data confirms thermal stability of the complexes. Various kinetic parameters have been calculated. The values of the thermal degradation of all the complexes are same. The negative values of the apparent entropy of activation indicate that the activated complexes have more ordered or more rigid structure as the reactions are slower than normal.
9. From XRD study indicates that metal complex (B_1Mn,) is of monoclinic crystal system.
10. The complexes are biologically active and exhibited enhanced antifungal and antibacterial activities as compared to their parent ligands.

From all the above concluding remarks, the metal complexes can be assigned following proposed structures.

Metal complexes of flavones B1, B2 R=H, -CH3

M= Mn(II), Co(II), Ni(II), Cu(II) and Cd(II)

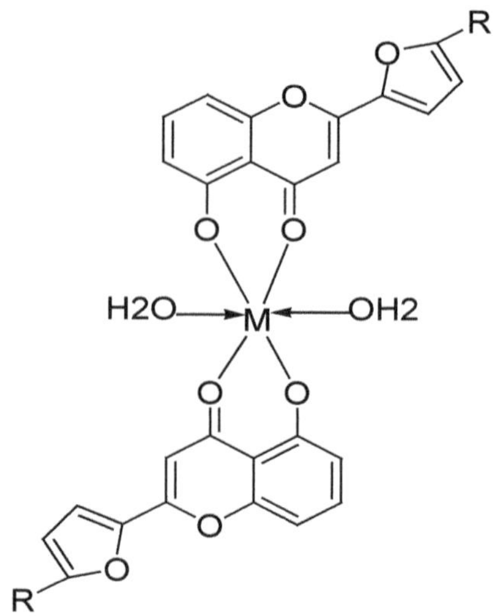

Metal Complexes of Flavones

Metal Complexes of Zinc Complexes

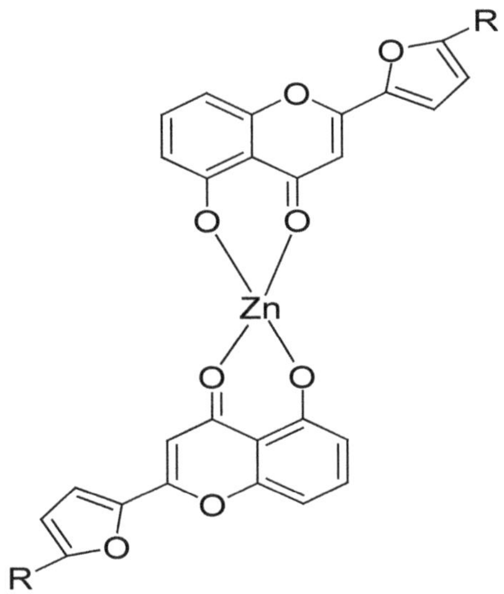

Fig. 5.1: Proposed Structures of Metal Complexes of Flavones.

References

Akaharia, T. (1928). Sci. Papers Inst. Phy. Chem. Res., 9, 165.

Ambrose, A.M. Robins, B.T. and Deeds F. (1950). Federation proceedings, 9,254.

Amoros M, C.O.M. Girre, L. Suavanger F. and Cornier M., *J. Nat.* (1992) *Prod.55*.1732.

Anderson, A.G. (2001). Journal of the American College of Nutrition, 20, 327- 336.

Azaroff, L. V. and Buerger, M. J. (1958). "The Powder Method in X-ray Crystallography "McGraw Hill, New York.

Azaroff, L.V., (1968). *Elements of X-ray Crystallography,* MC Grow Hill Book Company, New York.

Azoroff, L.V. (1960). *"Elements of crystallography"*, McGraw till Book Co. New York.

B.D. Cullity, (1978). Elements of X-ray diffraction Addison–Wesley, ISBN 0-201-01174-3 Chapter 14.

Bahad, P.J., Bhave, N.S., Aswar, A.S., (2000). *J. Indian Chem. Soc.77*, B 63.

Bamard, D.L. Simer, D.F. Hoffman, J.H. Meyerson, L.R. and Sidwell, R.W. (1993). Chemotherapy 39,203.

Barbucci,R., Caildi,G., Potecell,G.,Paoletti,P.,(1969). *J. Chem. Soc.,* (A) 1775.

Bassett, J., Denney, R. C., Jeffery, G. H., Mendham, J. (1978)., *"Vogel's Text book of Quantitative Inorganic Analysis"* IV Edn. ELBS and Longman Group Ltd. London, 342, 359-363.

Bassett, J., Denny, R.C., Jeffery, G. H., and Mendhan, J., (1977). "Vogels Text Book of Quantitative Inorganic Analysis", Heyden Philadelphia.

Baumann J. Von. Bruccham, Sen, F and Wurum, G.(1980). Prostglandins,20,627.

Beg, F. Bhatia, A.M. Baig,U.I. (2014 Jan/Feb). Biological activity of modified chrysin. *Int. J. of Multidisciplinary and Current Research,* 126-130.

Bhadeshia H.K.D.H. (2002). Thermal analyses techniques "Differential Thermal Analysis". University of Cambridge, Material Science and Metallurgy. www.msm.cam.ac.uk/phase-trans/2002/Thermal1.pdf.

Bhattacharya, K.C. (1934).*"An Elementary Physics for Indian School"*, The Indian press Ltd. Allahabad, 105.

Bhattacharya, K.C., (1934). "An Elementary Physics", The Indian Press Ltd., Allahabad Reva. Enld., 60-116.

Bunk,B.(2010). A short story about a big magic bug. *Bioengineered Bugs* **1**:85-91.

Burger, k., "Coordination Chemistry Experimental Method", I[st] Edn. Landon Buter Worth, 215.

Burger, M.J., (1960). *"Crystal Structure Analysis"* John Wiley and Sons, New York.

Carrasco, F., Thermochim.(1993). Acta, 213, 115.

Christen, Y. (2000) Journal of the American College of Nutrition, 71, 621–629.

Clayden, Greeves, Warren and Wother. (2001). *"Organic Chemistry"*, Oxford University, I Ed[n].

Clayden, Greeves, Warren And Wothers, (2002). "Organic Chemistry".

Coats, A., Redfern, J.P. (1964). Nature, 201, 68.

Collin, R.A. Schreiber, M. and Elvehjein, C.A. (1953). T. Nutrition 49,589.

Cullity, B.D. (1978). "Elements of X-ray Diffraction" 2nd Edition, Addison-Wesley Publishing Company Inc., London.

Davì G.A., Falco, C. (2005) Antioxid Redox Signal, 7, 256–268.

Demain, A.L., (1991). *Production of beta-lactum antibiotics and its regulation. Proceedings of the National Science Council Republic of China* Part B.,15,65 251.

Dharma Theja N, Tulsi Choudary P, Madhusudhan reddy I.(2011). International *Journal of Pharmacy and Pharmaceutical Sciences* vol. 3, suppl 2.

Dharwadkar, S.R., Karhanwala, M.D. (1969). Indian. J. Chem., 7, 729.

Donald L. Pavia, Gary M. Lampman, George S. Kriz, and James R. Vyvyan. (2015) "Introduction to Spectroscopy" V[th] Eds. Cengage Learning India Private. Ltd. ISBN 13:978-81-315-2916-4., 216-221.

Douglas, M., Douglass, F., David, J., Hamilton C., (1995). *J Hem. Soc.*, 269.

Edition, HoltSaundersCollegePublishers,NewYork.344-346.,484,713-726. *Voitovich, V. B.; Lavrenko, V. A.; Voitovich, R. F.; Golovko, E. I. (1994). "The* Effect *of Purity on High-Temperature Oxidation of Zirconium".* Oxidation of Metals. *42: 223–237*doi:10.1007/BF01052024.

Edn.

Elfin Susanti VH, Sabirin Matsjch and Tutik Wahyuningsih. (2014).Improved synthesis of 2,6-dihydroxy-3-4-dimethoxy chalcone by grinding technique to synthesize 5-hydroxy-3'-4'-dimethoxy flavones. *Indo. J. Chem.14* (2),174-178.3, 53-64 (1999).

Entrez Genome Project, NCBI

Erdogan A, Rao S.S. (2015 April). "Small intestinal fungal overgrowth". *Curr Gastroenterol Rep.* 17 (4):16.

European Molecular Biology Laboratory, European Bioinformatics Institute (EMBL EBI).

Eye R.W.M.D. Wait, E. (1960). "X-ray Powder Photography in Inorganic Chemistry" Academic Press, New York.

Faize Beg, Mrs. Amrita M. Bhatia, Mrs. Ulfat I. Baig. (Jan/Feb 2014). Biological activity of modified chrysin. *Int. J. of Multidisciplinary and Current Research,* 126-130.

Farhan, S. A. (2013). Study on the interaction of copper (II) complex of morin and its antimicrobial effect. *Int. J. Chem. Sci.:*11(3), 1247-1255.

Feldmann, Horst (2010). Yeast Molecular and Cell bio. Wiley-Blackwell. *ISBN 978-3527326099.*

Feng L; Senchenkova SN; Yang J; Shashkov AS; Tao J; Guo H; Zhao G; Knirel YA; Reeves P; Wang L (2004). "Structural and Genetic Characterization of the Shigella boydii Type 13 O Antigen". *Journal of Bacteriology.* 186 (2): 383–392.

Figgi's, B.N, and Lewis. (1964)., "The magnetic Properties of transition metal Complexes".

Figgi's, B.N, and Nylon, R.S, (1958). *J. Chem. Soc.,* 4190.

Figgi's, B.N. Lewis, J. (1964). Prog. Inorg. Chem. 5210.

Flynn, J.H., Thermal Analysis Schenker R.F. (Jr) and Gran P.D. (1971). Academic Press New York, 2, 1111.

Foster, J.W., (1949). *Chemical Activities of Fungi,* Academic Press, New York.

Freeman, E.S., Carroll, B.J. (1958). *J. Phy. Chem.* 62, 391.

Frost, A.A., Pearson, R.G. (1961). "Kinetics and Mechanism", John Wiley, New York.

Fuoss, R.A., Salyer, I.O., Wilson, H.S. (1964). *J. Polym. Sci.* Ra., 3147.

Gow, N.A.R. (2017). "Microbe Profile: Candida albicans: a shape-changing, Opportunistic pathogenic fungus of humans". *Microbiology.* 163 (8): 1145–1147.

Gupta, S. Kumar, N. Kumar S. Sharma, P.K. int. jou. Therapeutic application, Vol 7,1012,1-8.

Harborane, J.B. Mabry, T J. and Mabry, H.(1975). "The Flavonoids" Chapman and Hall London, 1011.

Hetland, G., Johnson, E., Lyberg, T., Bernardshaw, Tryggested, A., Grinde B., (2008). *Journal of Immunology,* 68, 4, 70-363.

Hitchon, C. El-Gabalawy H. (2004) Arthritis.

Hong-Yun, Z., Dong-Li, C., Pei-Kun, C., De-Ji, C., Guang-Xia, C., Hong-Quan,Z., (1992). Polyhedron, 11, 2313.

Horowitz, H.H., Metzger, G., (1994). Anal. Chem., 35, 7464.

Jadhav, S.B. Bagul, K.R. Bagul, P.R. Gaikwad, K.V. (2008) Oriental Journal of Chemistry, 24, 583-588.

Jagdale, A.N. (2006). *Ph.D. Thesis.* Dr. B.A.M.U. Aurangabad.

Jager, W. Zembsch, B. Wolschman, P. Pittenauer, E. and Senderarichh, A.M.(1998). Life, Sci; 62;186.

Jingliang Zhang, Chang Cui, Huili Zhang, Shiguang Wang, Jinhai Liu, Guangyu Zhai. Synthesis, characterization and antihyperlipidemic of rutin-calcium(II) Complex.

John, H.A., Teller, E., (1937). Proc. Roy. Soc.,A-161,220.

Kalinowski, J; Bathe, B; Bartels, D; Bischoff, N; Bott, M; et al. (4 September 2003). "The complete Corynebacterium glutamicum ATCC 13032 genome Sequence and its impact on the production of l-aspartate-derived amino acids and vitamins". *Journal of Biotechnology.* 104 (1–3): 5–25.

Kalsi, P.S.,(2002) "Spectroscopy of organic compounds," Vth Eds. New Age *Int. Pub. Ltd.* ISBN 81-224-1351-X., 7.

Karamunge, K.G. Vibhute, Y.B. (2012). Synthesis and structural studies of complexes of Cu(II), Ni(II) and Co(II) with 2-hydroxy chalcones; Int. J. Chem. Sci:10(1) ,490-500.

Kausar, H. Khan, M.U., Nuzhat S.K Begum, Baig, MMV and Baseer M.A.(2013). Synthesis of some new halogen substituted flavones derivatives. *International Journal of Phytopharmacy Research*, vol 5 issue 2, 76-79.

Kausar, H. Khan, M.U., Nuzhat, S.K., Begum, Baig, MMV and Baseer, M.A. (2013). Synthesis of some new halogen substituted flavones derivatives. International Journal of Phytopharmacy Research, vol 5 issue 2, 76-79.

Kemp, W., (1991). "Organic Spectroscopy", MC Millan Press IInd Eds.

Kerawala C, Newlands C, eds. (2010). Oral and maxillofacial surgery. Oxford: Oxford University Press. 446, 447. ISBN-0-19-920483-0.

Kittel, C., (1994). "Introduction to Solid State Physics" Vth Edn. Wiley Eastern Ltd-New Delhi.

Klug,H.P.Alexander,L.E.(1974)."X-ray Diffraction Procedures for Polycrystalline and Amorphous Materials" 2nd Edition, Jhon Wiley and Sons, New York.

Koes, R.E. Quattrocchio, F. (1994). The flavonoid biosynthetic pathway in plants Function and Evolution BioEssays,16,123-132.

Kumari S. A., Babu B. K., Muralikrishana, N., Rao K. M., Sinduri V.L.,Praveen K., Rao V. P., Latha, B. S., Ramrao, K., Veeraiah V., and Kota Praveen. (2015). Riboflavin metal complexes. *Dar Pharma Chemica*, 7(2). 307-315.

Lachman, J. Dudjak, J. Kolihova, D. Pivec V. (2005). Effect of cadmium on flavonoid content in young barley plants. *Plant Soil Environ*. 51, (11):513-516.

Lachman, J. Dudjak, J. Miholova, D. Kolihova, D. Pivec V. (2005). Effect of cadmium on flavonoid content in young barley plants. *Plant Soil.Environ*, 51, (11):513-516.

Lackmann, G.M. Claeys, M. Rwangabo, P.C. Herman, A. Gand Vlietinck,A.J. (1996). *Plant. Med.*, 52,433.

Lee, J.D., (1996). "Concise Inorganic Chemistry", V$^{th\ Eds}$, 198-199.

Lever, A.B., (1968). "Inorg. Electronic Spectroscopy" Elservier, 356-559.

Liohr, A.D. (1960). J. Phys. Chem., 64, 43.

Lipson, H. and Steeple, H. (1970). "Interpretation of "X-ray Powder Diffraction Pattern" MacMillan, London.

"Magnetizability, ξ".*(1997). IUPAC Compendium of Chemical Terminology-The Gold Book (2nd ed.). International Union of Pure and Applied Chemistry.*

Mac Callum, J.P, Tanner, J. (1970). Eur. Polym, 6, 1033.

Madan, M. and Thind, K.S., (1998). *Physiology of Fungi,* A. P. H. publishing corporation.

Mahdi Al-Hassani, R.A. Nife, O.A. Mahmood, A.T. Awad, A.A. (2016). Synthesis, characterization theoretical studies and biological activities of naringin metal complexes; ICAIJ, 11(2), 047-057.

Mahmood, N. Pizza, C. Aquino, R. De. Tammasi, N. Piacente, S. Colman, S. Bruke, A. and Hay A. J. (1993). *Antiviral. Res.*22, 189.

Mahmoud Najim-Al-Jibouri and Musa T.M. (2017). Synthesis and spectroscopic study of Cobalt(II), Nickel(II), Copper(II) and Zinc(II) complexes derived from Bi-dentate-2-pyrazoline based ligand. *Int. J. Chem Sci;*15(2):140.

Mane, P.S., Saluke, S.D., Shiradkar, S.G., Chondheker, T.K. (2001). *Asian. J. Chem.,* 13, 1573-1576.

Martins N, Ferreira IC, Barros L, Silva S, Henriques M (2014 June). "Candidiasis: predisposing factors, prevention, diagnosis and alternative treatment". *Mycopathologia.* 177 (5–6): 223–240.

Mathews, P.,(2002). "Advanced Chemistry" Chembridge Uni. Press.,128-129.

Mathews, R.W., Walton, R.A. (1971). Inorg. Chem.,10, 1433.

Metha, Akul (2011 Dec13). "Principle" Pharma X Change.info.

Metha, Akul (2012 Apr 22). "Derivation of Beer–Lambert Law". *Pharma X Change. info.*

Metler Toledo, (2012). Reducing Common Errors in Conductivity measurement.

Mikkola, Raimo. Food and Indoor Air Isolated Bacillus Non-Protein Toxins:Structures, Physico-Chemical Properties and Mechanisms of Effects on Eukaryotic Cells (PDF). 12. Retrieved 2015-10-24.

Miller, A.L. (1996) Alternative Medicine Review,1, 103-111.

Mira Madan and Thind, K.S. (1998). *Physiology of fungi,* A. P. H. Publishing Corporation.

Mishra, A.P., Khare, M.G. (1998). *Indian J.Chem.Soc.,* 77, 10, 927.

Mishra, A.P.,Khare, M.G., (1998). *Indian J. Chem.,* 37A, 10, 927.

Mohamad, G.G., Nour-El Dien, F.A., El-Gamel, E.A. (2002). J. Therm. Anal. Cal., 67,135.

Mokle,S.S. Khansole, S.V. Patil, R.B. and Vibhute Y.B.(2010).Synthesis and Antibacterial activity of some new chalcones having 2-chloro-8-methoxyquinolinyl moiety; *International Journal of Pharma an Biosciences* V_1(1).

Molina J.A.E. and Alexander M. Can, T. (1967). Microbiology, 13,819.

Morand, C. Crespy, V. Manach, C. Besson, C. Demigne, C. Remesy, C. (1998). Am. J. Physiol. 275, R212-R-219.

Munde, A.S. (2007-Dec). *Thesis Ph.D.* Dr. B. A. M. U. Aurangabad.

Murti Y. and Mishra P. (2014 March/April). Synthesis and evaluation of flavonones as anticancer agents. *Indian Journal of Pharmaceutical Sciences.*

Naclerio, G; Ricca, E; Sacco, M; De Felice, M(1993-December). "Antimicrobial activity of a newly identified bacteriocin of Bacillus cereus". *Appl Environ Microbiol.* 59 (12): 4313–6. PMC195902 PMID 8285719.

Naghski, J. Copley, M.J. and Couch, J.F. (1947). Science, 105,125.

Nagwa Nawar, Nassir, M.H. (2000). Transition Metal Chemistry, 25, 1-8.

Norman, F.M., Henry and Kathiean Lansdale, (1969)."International Tables for X-ray Crystallography", IIIrd Edn, Kynoch Press, Birmingham.

Nunomura, A.R. Castellani, X.Zhu, P. Moreira, G. Perry, M. S. (2006) Journal of Neuropathology & Experimental Neurology, 65, 631–641.

Odds, F.C. (1988). Candida and Candidosis: A Review and Bibliography (2nd ed.). London; Philadelphia: Bailliere Tindall. ISBN978-070201256.

Oggioni, M.R., Pozzi, G., Valensin, P.E., Galieni, P., Bigazzi, C., (1998). Recurrent septicemia in an immunocompromised patient due to probiotic strains of *Bacillussubtilis". J. Clin. Microbiol.* 36, 325–6. PMC 124869 PMID 9431982.

Omar H. Al-Obaidi, (2014). *Journal of Applied Chemistry,* 3(1) 157-163.

Omar, H. Al-Obaidi. (2014). Synthesis, spectral characterization, theoretical evaluation of new Co(II) and Mn(II) complexes of flavones., *Journal of Applicable Chemistry.* 3(1):157-163.

Omar. H. Al-Obaidi (2014). Synthesis, spectral characterization, theoreticalevaluation of new Co(II) and Mn(II) complexes of flavones., *Journal of Applicable Chemistry,* 3(1),157-163.

Omar. H. Al-Obaidi (2014-Aug.). Study of the cytotoxicity effect of new Co(II), Mn(II) Ni(II), and Cu(II) complexes of chalcone on cancer and Antimicrobial activity, PCE Vol 1, Issue 2, 14-18.

Omar. H. Al-Obaidi, (2014). Synthesis, spectral characterization, theoretical evalution of new Co(II) and Mn(II) complexes of flavones., *Journal of Applicable Chemistry,* 3(1):157-163.

Omar. H. Al-Obaidi. (2014). Synthesis, spectral characterization, theoretical evalution of new Co(II) and Mn(II) complexes of flavones., *Journal of Applicable Chemistry.*3(1):157-163.

Omar. H. Al-Obaidi. (2014). Synthesis, spectral characterization, theoretical evaluation of new Co(II) and Mn(II) complexes of flavones., *Journal of Applicable Chemistry* 3(1):157-163.

Omar. H.Al-Obaidi. (2014).Synthesis, spectral characterization, theoretical evaluation of new Co(II) and Mn(II) complexes of flavones. Journal of Applicable Chemistry,3(1):157-163.

Orgel, L.E. (1966). *"An introduction to Transition Metal Chemistry"* Wiley IInd Edn., 5, 7.

Orgel, L.E., (1966). "An Introduction to Transition Metal Chemistry" Wiley IInd

Ozdemir, A. Zitouni, G.T. Reveal, K.G. and Guven, K. (2007). "Synthesis and antimicrobial activity of 1-(4-aryl-2-thiazolyl)-3-(2-thienyl)-5-aryl-2-pyrazoline derivatives," *Eur. J. of Med. Chem*, vol.42, issue 3, 403–409.

Panhwar, Q.K. Memon, S. (2014). Synthesis, characterization and antioxidant activity of rutin complexes. *Pak. J. Anal. Environ. Chem.* Vol.15.

Patharkar, V.R. (2010). Ph.D. Thisis, Dr.B.A.M.U. Aurngabad. 156-157.

Patherkar, V. R., (2010). *Ph. D Thesis*, Dr. B. A. M. University, Aurangabad.

Patil,S.A.Naik,V.H., Kukarni,A.D., Badami,P.S.,(2010).*J.Sulphur.Chem.*, 31, 109

Paula, Peter Atkins, Julio de (2009). Elements of physical chemistry V[th] ed. Oxford: Oxford Uni.Press., 459. ISBN 978-0-19-922672-6.

Piloyan,G.O., Novikova,O.S. (1966). *Russ. J. Inorg. Chem.*,12, 313.

Pope, M.D.,(1977).*"Differential Thermal Analysis"*, Heyden, Philadelphia.

Pratt D.E. (1992). ACS Symp. Ser. No 507, 54-71.

Prez, A.R., Abanes -Demello, A., Poglano, K., (2000). Spo IIB Localizes to to Active Sites of Septal Biogenesis and Spatially Regulates Septal Thinning during Engulfment in *Bacillus subtilis. Journal of Bacteriology.* February;182:1096-1108.

Rajbhoj, S.W. and Chondhekar, T.K. (2000). "Systematic Experimental Physical Chemistry", Anjali Publication Aurangabad. IInd Eds, 304-315.

Rao, C.N.R., (1973)."University General Chemistry" An Introduction to Chemical Science Mc. Millan Indian I[st] Edn., 268.

Raper, K. B., Thomas, C., (1945). A manual of Aspergillus IX,.Williams and.Wilkins Company (Baltimore), 76, 3-37.

Rathore, M.M., Rajput, P.R., Parhate, V.V. (2015-Jan). Synthesis and antimicrobial activity of some chalcones and flavones. *International Journal of Chemical and Physical Sciences*, Vol 4 special issue NCSC.

Raulin, J., (1969). *Ann. Sci.*, Nat V. Botan 11, 92.

Reddy, K.V., (1998). "Symmetry and spectroscopy of molecules" IInd Eds. New Age Int. Pub. Ltd., 551.

Reddy, Vidyavati. Patil, N. and Angadi S.D. (2008 July). Synthesis, characterization and antimicrobial activity of Cu(II), Co(II) and Ni(II) complexes with O,N and S doner ligands; *E-Journal of Chemistry* Vol. 5, No.3, 577-583.

Rice-Evans, C.A. Miller, N.J. (1996). Biochem. Soc Trans.,24,790-795.

Rodriguez-Fernandez, E.,Garcia, E.,Hermosa, M.R., Limenez-Sancher, A., Ssnchez, M.M., Monte,E., Criado, J.J. (1999). *J. Inorg. Biochem.*, 75, 181

Roger Grinter, (2005). The Quantum in Chemistry: An Experimentalist's View, John Wiley & Sons, ISBN 0470017627, 364.

Rout, S.S. Sahoo, R.N. Pattanaik, S.P. Pal, A. Sudam Chandra SI, Mohanty P. (2013 April). Anti-nociceptive activities of complexes of Naringin with Co(II) metal ions. *International Journal of Pharmacy and Pharmaceutical Sciences.*, vol.5 issue3,20.

Ryan, K.J.,Ray,C.G.(2004).*Sherris Medical Microbiology,*IVthEd.(Mc Graw Hill).

Ryan, Kenneth James; Ray, C. George, eds. (2004). Sherris medical Microbiology: an introduction to infectious diseases (4 ed.). McGraw-Hill Professional Med/Tech. ISBN 978-0-8385-8529-0.

Sacconi, L. (1960). Transition Met. Chem., 64, 43.

Saini, P.P., Kumar,V., Gupta, A.K. Gupta, G K., (2014). Met. Chem. Res., Gt 23, 690-698.

Schaechter, M., Ingraham, J.L., Neidhardt F.C., (2006). *Microbe.* (ASM Press, Washington, DC.

Sekerei, M., Yakuphanoglu, F., J. (2004). Therm. Anal. Cal., 75, 189.

Sengupta, S. K., Pandey, B. K., Shrivastava, B. K., and Sharma, V. K.,(1998). *Transition Metal Chem.*, 23, 4, 349-353.

Shah, N; Sattar, A; Benanti, M; Hollander, S; Cheuck, L (2006 January). "Magnetic resonance spectroscopy as an imaging tool for cancer: a review of the literature". *The Journal of the American Osteopathic Association. 106 (1): 23- 27.* PMID16428685

Sharma, Y.R. (1980) "Elementary Organic Spectroscopy" Ist Edn. S. Chand Company Ltd. ISBN 81-219-0919-8., 137-207.

Sharp, J.B. Wentworth, S.A. (1969). *Anal. Chem.,* 41, 2060.

Shastri, R.A. (2015 4thJune). Synthesis and antimicrobial activity of some novel pyrazoline and flavones derivatives derived from furan chalcone; *World Journal of Pharmaceutical Research,* Vol.4, issue7, 642-651.

Shastri, R.A. (4 June 2015). Synthesis and antimicrobial activity of some novel pyrazoline and flavones derivatives derived from furan chalcone; *World Journal of pharmaceutical Research,* vol 4, issue 7, 642-65.

Shirodkar, S.G., (1997). Ph.D. Thesis, Dr. B.A.M.U. Aurangabad.

Shirodkar,S.G. (1997). *Ph.D Thesis,* Dr. B.A.M.U. Aurangabad.

Silverstein, R.M., Webster F.X. (2002). "Spectrometric Identification of Organic Compounds" VIth Eds. John Wiley and Son, Inc.72-73.

Singh, R.V., Chaudhari, A.,Joshi,S.C., (2004). *Main- Group Metal Chem.*, 27, 59.

Skinner, C.E., Emmiens, C.W., and Tsuchiya, H.M., (1947). *Molds yeast and Actinomycetes,* John Wiley and Sons, New York, IInd Edn, XIV 136, 409.

Skoog, D.A., (1985). "Principles of Instrumental Analysis", IIIrd International

Stefanini I, Dapporto L, Legras JL, Calabretta A, Di Paola M, De Filippo C, Viola R, Capretti P, Polsinelli M, Turillazzi S, Cavalieri D (2012). "Role of Social wasps in Saccharomyces cerevisiae ecology and evolution". Proc. Natl. Acad. Sci. U.S.A.109(33):13398.403.Bibcode:2012PNAS.10913398S.doi:10.1073/pnas.1208362109. PMC 3421210. PMID 22847440.

Sutton, D. (1968). "Electronic Spectra of Transition Metal Complexes, "MC. Graw Hill, 145-178.

Tatsumi, Nami; Inui, Masayuki (2012,14-August). Corynebacterium glutamicum: Biology and Biotechnology. Springer Science & Business Media. 336. ISBN 978-3-642-29857-8.

Tereschuk, M.L. Riera, M.V.Q. Casteo, G.K. and Afdala, L.R.J. (1997). *Ethano Pharmacol* 56,227.

Thenmozhi Muthurajan; Evaluation of Quercetin complex as an efficient positive contrast enhancer for magnetic resonance imaging; Royal Society of Chemistry, www.rsc.org/advances ,1-34.

Thom, C., and Raper, K.B., (1945) Amunal of Aspergillus XII 13-73 Willams and Wilkins Co Baltimore, 76.

Turnbull PCB (1996). Baron S; et al., eds. Bacillus. **In**: Baron's Medical Microbiology (4th ed.). Univ of Texas Medical Branch.ISBN 978-0-9631172-1-2.

Under Kofler, L.A., and Hickey, R.J., (1954). *Indusrial Fermentation*, Chem. Pub. Co., New York.

Valentina Uivarosi, Mihaela Badea, Rodica Olar, Bruno Stefan Velescu, Victoria Aldea. (2016). Synthesis and characterization of a new complexes Off Oxovanadium (IV) with Naringenin, as potential insulinomimetic agent, FARMACIA, vol. 64.

Varma, S.D. and Kinoshita, J.H. (1976). *Biochem Pharmacol.* 25, 2505.

Venkatachalam, H. Jayashree, B.S. Nigam, S. (2017 Sep-Oct). A preliminary cytotoxic study on the zinc complexes of flavonols.; *Indian Journal of Pharmaceutical Sciences.*, 838-843.

Venkatachalam, H.,Jayashree, B.S. Nigam, S. (2017 Sep-Oct). A preliminarycytotoxic study on the zinc complexes of flavonols. *Indian Journal of Pharmaceutical Sciences*. 838-843.

Wendlat, W.W.,(1977). "Thermal Methods of Analysis", Heyden Philadelphia.

Williams, D.H., and Fleming, I.,(1973). "Spectroscopic Methods in Organic Chemistry", IIIrd Edn. London MC Graw.

Wood-Kaczmar, A.S. Gandhi, N. (2006) Molecular Medicine, 12, 521–528.

Xavier, A. Gobu, P. Srividhya, N. (2005 May). Schiff base and its transition metal Complexes. *International Journal of Multidisciplinary Research and Development.* vol. 2, issue 5,134-138.

Yuanzhen Liu, Mingquan Guo (2015). Studies on transition metal quercetin complexes using Electrospray Ionization Tandem Mass Spectroscopy; Molecules 20, 8583-8594, doi: 10.3390/molecules20058583.

Zahoor A; Lindner SN; Wendisch VF (2012-October). "Metabolic Engineering of Corynebacterium glutamicum Aimed at Alternative Carbon Sources and New Products". *Computational and Structural Biotechnology Journal*. 3 (4): 1–11.

Zorgo E, Chwialkowska K, Gjuvsland A.B, Garré E, Sunnerhagen P, Liti G, Blomberg A, Omholt S.W, Warringer J (2013). "Ancient evolutionary trade-offs between. yeast. ploidystates". PLoS Genet. 9(3):e1003388.*doi:10.1371/journal.pgen.1003388. PMC3605057. PMID 23555297.*

Index

www.ingramcontent.com/pod-product-compliance
Ingram Content Group UK Ltd.
Pitfield, Milton Keynes, MK11 3LW, UK
UKHW041306190726
13845UKWH00001B/5

9 789354 616020